ALSO BY PETER D. WARD AND DONALD BROWNLEE

*Rare Earth*


ALSO BY PETER D. WARD

*Future Evolution*

*Rivers in Time*

*Time Machines*

*The Call of Distant Mammoths*

*The End of Evolution*

*On Methuselah's Trail*

*The Natural History of Nautilus*

*In Search of Nautilus*

# the life *and* death *of* planet earth

# the life *and* death *of* planet earth

### How the new science of astrobiology charts the ultimate fate of our world

## PETER WARD *and* DONALD BROWNLEE

**PIATKUS**

Copyright © 2002 by Peter D. Ward and Donald Brownlee

First published in the USA in 2002 by
Times Books, Henry Holt and Co., LLC

First published in the UK in 2003 by
Judy Piatkus (Publishers) Limited
5 Windmill Street
London W1T 2JA
e-mail: info@piatkus.co.uk

**The moral right of the author has been asserted**

*A catalogue record for this book is available
from the British Library*

ISBN 0 7499 2425 X

This book has been printed on paper manufactured
with respect for the environment using wood from
managed sustainable resources

Printed and bound in Great Britain by
Mackays of Chatham Ltd, Chatham, Kent

*For Patrick, Allison, Carson, and Nicholas*

## Fire and Ice

Some say the world will end in fire,
Some say in ice.
From what I've tasted of desire
I hold with those who favor fire.
But if it had to perish twice,
I think I know enough of hate
To know that for destruction ice
Is also great
And would suffice.

—ROBERT FROST

# ·CONTENTS·

# the life *and* death *of* planet earth

# PROLOGUE

FIRE? OR ICE?

Come with us to the future of Earth, a world that echoes our prehistoric past.

Imagine our planet some tens of thousands of years into the future, a stretch of history far longer than the time it has taken our species to develop from hunter-gatherers to industrial civilization. From the vantage point of a derelict and forgotten satellite orbiting far out in space, the reflection of our marbled home is as disquieting as it is dazzling: a reflective, expanding white. The ice of the Poles is creeping steadily equatorward as glaciers advance, and the snows of winter are persisting far longer into the increasingly brief summers. The Alps, the Himalayas, and the northern Rockies are capped year-round with growing tongues of ice. Even at Mount Kilimanjaro and the Mountains of the Moon in central Africa, the glaciers are growing. The sea level that briefly rose at the height of civilization is now dropping, exposing new coastal plains, linking islands, and creating land bridges. Harbors have become meadows.

The English Channel and the Bering Strait have become corridors. All the maps have changed.

At night the planet no longer glistens with a galaxy of city lights that once stretched from the Arctic to the Southern Ocean. Instead, the Arctic has been abandoned and the Southern Ocean is largely frozen over. The lights that glitter are in a narrower band hugging the equator and midlatitudes. Many are now campfires.

It is as if time has not gone forward, but backward. Eerily, the planet is beginning to resemble again the Ice Age world that our primitive ancestors endured.

The age of fossil fuels is long over, the planet's reservoir of oil and gas and coal expended in a gluttonous feast of energy consumption that briefly dumped billions of tons of carbon dioxide into the atmosphere. The resulting global warming caused agricultural havoc and erratic climate swings that lasted for several disastrous centuries, but that's but a blink in planetary time. Slowly, the natural processes that seek to balance our planet reabsorbed the carbon out of the atmosphere. The cruelly hot weather dissipated, and for a while our species rejoiced at a return to "normality." But now the climate has dipped toward a more ominous norm. Earth is returning to the conditions that have dominated it for the last 3 million years: a regime of ice. The human civilization that arose in a brief interglacial period is now struggling for survival in a colder and much drier world. The diverse and extensive rain forests of the tropics are being replaced by savanna. The midlatitude grasslands that once helped feed the world, such as the American Great Plains, are becoming dust bowls. Katabatic winds that can reach 200 kilometers per hour howl off the advancing ice sheet and make permanent habitation near its fringe almost impossible. The glaciers are a blue-white wall, gritty on top and at their base, which are grinding forests, towns, and highways into oblivion. Eventually the skyscrapers of Manhattan and the towers of London will be bulldozed by snouts of ice half a kilometer high. Earth has become a planet where humans struggle to feed themselves. Changing climate has made a mockery out of seasons, and the farm crops that predictable seasons allow. Our descendants are starving.

Shiver, and go on.

Come with us to an even grimmer future. The reign of ice will come again, but will not last forever. Fire will succeed it, in the form of an increasingly hot sun.

This time we travel not just thousands of years forward, but hundreds of millions of years: a time more distant from the present era than the ancient seas and primordial swamps that preceded the dinosaurs. The succession of Ice Ages that held the Earth in thrall for millions of years is long forgotten. From space, our planet no longer looks white, or green, or even blue. Its continents are a desolate reddish-brown and its atmosphere thick with windblown dust. Descend in your imagination to an alien world.

Picture that we are standing on a seashore, at the edge of a vast, white-capped ocean. The water, at least, is familiar. As it has been since nearly the dawn of time on Earth, the sea is filled with life. Fleet schools of fish arrow through the sunlit surface waters. Below, on the rippled sand of the sea bottom, persist crabs and anemones, flounders and starfish, corals and barnacles. What is different is that the ocean—once a cradle for the life that crawled onto land—has become a sanctuary. Animals, hammered by a relentless sun, are retreating into the water.

The ocean shore that was once life's beachhead has become its Dunkirk, and the species that cannot readapt to the water are doomed to extinction. The sandy strand we are standing on has become an oven where a hard cadre of animals struggles to exist between the two worlds of warm water and far warmer air. During high tide a few species of hardy crustaceans and mollusks scurry and hunt, feeding and breeding. But during low tide all visible life comes to a stop, the animals hunkered down under parasol-like shells or wedged within the damp crevices of overhanging rocks, trying to survive the murderous rays of the sun.

Look up. The sky is grayish-yellow, huge winds carrying storms of sand at galelike velocities. The continents have become deserts of scoured rock and marching dunes. Although imperceptible to the eye, the Sun is slightly brighter and the Moon—slowly spiraling away—appears slightly smaller and dimmer. The hottest temperature on record in our own times is the 136 degrees Fahrenheit measured at Le Aziza, Libya, on September 13, 1922. In this future world it reaches that temperature every day, and not just in the Sahara but on midlatitude shores that were once cold and forested.

The humidity is 90 percent, the air sticky, oxygen thin. We gasp for breath as if on a high mountain.

What life is left? There is no driftwood on this beach, because there are no longer any trees on Earth. Or bushes. Or even grass. The tallest green things left are mosses, found among the more common lichens and fungi that cling to a precarious existence. Even soil is a thing of the past, for when the dying roots of the planet's flora unclenched their grasp on the topsoil, it flew with the wind, leaving behind rock, dust bowl, and dune. The land has become a vast expanse of sand and rock. The rivers are chocolate Colorados carrying eroded land to sea.

Some animals persist in this new hell. If we get down on our knees we can see that centipedes and spiders still prowl for insect prey. Ants march in search of wrack worth scavenging. Amphibious lizards watch for food. All of these species conduct their business with alacrity, however, scurrying frantically to finish before the punishing sun reaches its awful noon. Then they too must hide.

There are no more birds. Nor mammals, nor amphibians. There is no song and no shade. To try to escape the heat, we too wade into the water, but the surf is uncomfortably hot. We'd have to develop gills and dive deeper to find temperatures that are more comfortable. Yet even the sea will turn out to be only temporary sanctuary for the animals that persist there. The vast oceans themselves are evaporating, their water molecules slowly lost to space.

Even they, like all things, will eventually come to an end. The long history of the sea will have left only glistening plains of salt.

HOW DO WE KNOW SUCH THINGS? IN AN AGE WHEN WEATHER forecasting is reliable only a few days into the future, how can scientists predict the fate of our planet millions or even billions of years into the future?

Our confidence is very recent. It is only in the past few decades that science has begun to adequately understand the full life cycle of planets, and thus the probable life and death of planet Earth. Geology, oceanography, biology, atmospheric science, and astronomy are just some of the disciplines that have contributed to discoveries about how our planet works, how it will respond to a relentlessly brightening Sun, and how it will ultimately break down. Our neighboring planets Venus and Mars, one blisteringly hot and the other frozen, have provided valuable insights into how rare, unique, and wonderful our own home is. They also post grim warning that the nurturing environment we take for granted can change. Will it? For the first time in history, a species has arisen that has the power to utterly modify its home world, for the worse or for the better. That species is us.

Like Renaissance doctors probing the wonders of the human body and mapping entirely new systems like blood circulation, scientists today are beginning to understand the cycles that make Earth work. As early as the eighteenth century, one of the pioneers of geology, James Hutton, applied the metaphor of circulation to his new science, suggesting that rock circulates like blood in an endless cycle of creation, erosion, and subduction. Hutton's metaphor is more accurate than he ever imagined. As strange as it sounds, our air and water are kept in balance by this circulation of rock, as we shall explain in the chapters to come. It will be the drift and then freeze-up of rock, coupled with changes in the Sun, that will upset this balance.

This science of how planets live and die has a new name, astro-biology, and it has combined the discoveries of space probes and astronomers with the discoveries of Earth scientists. We don't have just one world to study when seeking to understand how nature works, we have a solar system of worlds. We don't have a single sun to ponder, but countless suns sprinkled throughout our galaxy that are in all the various stages of birth, vigor, decline, and death. And we don't just have our own brief moment in time to learn from, we have an increasingly sophisticated understanding of natural history that extends back hundreds of millions and even billions of years. The Space Age, coupled with discoveries into Earth's distant past, has helped cause a quiet revolution in our understanding of our own planet that is as dramatic, in its own way, as the discovery of the New World.

Astrobiology is a discipline that is useful for predicting the kinds of planets that could harbor life, to search for such life, and to understand the life cycle of our own world. The authors of this book are astrobiologists and our thesis is simple: Earth's future—and its perilous present—can be understood for the first time by studying its past and our neighbors in space.

Our science's methodology is explained by its new name. The "astro" part is pretty self-explanatory, coming from a very old and traditional field of science: astronomy. Astronomers are already well positioned to study the end of our world, since they have long studied the ends of other worlds far beyond the Earth. In their powerful telescopes, astronomers routinely observe objects in deep space that surely are the destroyed remnants of once habitable solar systems. They study the nebula left behind from supernova, or exploding stars. They analyze the energetic burst of black holes burping out enormous gouts of lethal gamma rays, as whole stars are swallowed into their yawning gullets. They study literally millions of stars to develop a physical understanding of how stars work and evolve. This work makes it possible to predict the behavior of stars, including our own, billions of years into the future.

Astronomers also detect the presence of Jupiter-sized planets in highly eccentric orbits close to their stars, the sudden flare of variable stars, and even spectacularly odd galaxies that are formed by the collision of two spiral galaxies like our own. These astronomical insights give luminous testament to the end of other Earths: other planets that had water, warmth, life, and perhaps even other intelligent races extirpated by cosmic catastrophe. Yet all of the astronomical data come to us as light emitted long ago, from fantastic distances away. They are impersonal, antiseptic, and until recently seemed to give no clear lessons about the details on how our own Earth and its life might end. Astronomy can tell us only that other worlds might have ended.

To complete our understanding we must turn to the other half of our new name, "biology." To understand how life must end, we must understand how it arose, how it modified the physical features of our planet, and how it persists. Much of this knowledge comes from our growing understanding of contemporary plants and animals. And much else comes from paleontology, the study of life in the past and its extinctions. Paleontology can tell us about worlds of life on this planet that have already ended, and how those ends came about. The death of the dinosaurs is the most famous of these endings, but there were other extinctions even greater than that one. These ends are left as signposts in the Earth's rock record, and can be readily observed at many places on Earth.

Astronomy, geology, and paleontology are sciences of the past. The rocks, fossils, and even the light we receive in telescopes are but bits of the past brought into the present day. To understand the future, we need more than this, however. We must understand how our present world works, and construct "models" of it made not of clay and balsa wood, but of numbers, physics, and chemistry. These mathematical models—based on the universal laws of nature—become possible futures, and we test their accuracy by comparing their projections to the real Earth, and real planets, of today. Teams of scientists around the world have constructed such models to examine what a change in sunlight or geography or atmosphere might do to climate, ocean currents, oxygen levels, or other variables. We have pulled their findings together to write this book.

Two variables, above all others, will control the destiny not only of life on Earth, but also, ultimately, of the planet itself. The first is the amount of carbon dioxide—the nutrient of life on the planet—which resides in water and in the atmosphere. This greenhouse gas helps regulate our world's temperature, and our planet is extraordinary in the systems it has to balance that greenhouse gas with incoming radiation. The second, and ultimately far more important variable, is the amount of solar energy hitting our planet. The Sun's output is changing over time. Knowledge of the physical processes in stars allows us to predict what these levels will be in the far future, and what their effects will be.

Everyone knows how fallible weather prediction is, of course. So how can anyone possibly suggest that there can be any sort of confidence in climate prediction that will occur thousands to millions of years into the future? It turns out that future climate prediction is actually easier—and probably more reliable—than the weather forecasts we use in our daily lives. For the short term we want to know the expected temperature, the possibility of rainfall, and the possibility of dangerous weather such as thunderstorms or tornadoes. But predicting a brief weather pattern at a specific time and place is difficult because there are so many variables. Weather is a chaotic system in which a tiny change can create a domino effect of consequences to the point where reliable prediction is impossible more than a few days from now.

But predicting truly long-term climate change is easier because we are not so concerned with details like what the temperature will be on Sunday afternoon but rather the general state of climate at a general time. The level of success in predicting long-term planetary change should be much better than when dealing with the details of the present. Planets are highly complex bodies and—like people—their course of action is influenced by many different and often random factors. Their day-to-day changes may seem chaotic but their final states are highly predictable. Like a meandering stream, the precise pathway at a given time may not be predictable but it is certain that the flow will be downhill and that the stream will ultimately reach the sea.

Our book began as a quest to predict the scientific, physical end of the world. While researching it we realized, somewhat to our surprise, that there will not be one "end of the world" but many: a succession of dramatic shifts in climate, biology, landform, and atmosphere that will make our Earth as bizarre and alien a place in the distant future as it was in its distant past. To our knowledge, this book is the first comprehensive presentation of these "ends," based on our new understanding of planetary and stellar evolution. It attempts to pull together the myriad recent discoveries that suggest how our planet began, lives, and will die.

We believe that no task is more important at the dawn of the twenty-first century. Historians warn that those who fail to study the mistakes of the past are doomed to repeat them. We think that a failure to understand Earth's unusually rare history and probable future will leave us misunderstanding how wondrous, fragile, and perilous our present world—the world we think of as "normal"— really is. Our goal is not just dire prophecy of inevitable doom. It is to add to the thinking about the stewardship of our present-day world, and to start a discussion of the long-term future of our planet and our species. In the full spirit of the new science of astrobiology, it is also to understand our place and time in the cosmos.

· 1 ·

# OUR BLINK OF TIME

WE LIVE IN A TURBULENT PERIOD OF HUMAN HISTORY, A TIME of catastrophic wars, sweeping political movements, revolutionary social change, bewildering discovery, and religious and philosophic tumult. Never before has human population, technology, and information grown with the speed we see today. Never before has the pace of scientific discovery been so rapid, and the synthesis and absorption of that discovery so difficult.

Many of us periodically retreat from this frenzy by seeking refuge with nature in order to find some constancy in our lives. The natural world, at least, seems reassuringly stable. We assume the Sun will come up tomorrow, as the Earth reliably rotates on its axis. We count on the cycle of the seasons, as our planet orbits its Sun. We publish booklets that predict the tides a year in advance. We confidently await the annual migration of birds or salmon, and we plant gardens on expectations of rain, sunshine, and temperature. It is human nature to regard the world we are used to—the wide blue oceans, lush tropics, wind-lashed prairies, and snowcapped mountains—as a permanent thing. When we seek to "preserve" nature, the unspoken assumption is that the wilderness we are protecting will be reassuringly unchanging.

From the perspective of our own brief lifetimes, of course, this expectation of an unchanged natural world can sometimes be met. Yet a central argument of this book is that the home planet we regard as normal is anything but. In our previous book, *Rare Earth,* we argued that contrary to the dreams of science fiction authors, habitable planets with intelligent life such as our own might be extremely rare in the Universe. In this book, we will argue that even the Earth that we take for granted today is a temporary phenomena: that civilization has arisen in a rare respite from the Ice Age and an odd period of climatic stability, and that in the distant future our planet—unless we change or escape it—will be as bizarre and hostile a place to humans in the future as it was in its beginnings. Just as plant and animal life grows, thrives, dies, and decays, our planet is destined for a similar cycle. In a sense, history will begin to run backward as Earth's environment eventually slips toward the simpler ecology of hundreds of millions of years ago. This decline, we assert, is not just coming, it has already started. Biologically, Earth has already peaked—perhaps as long as 300 million years ago—and we are already living in a relatively impoverished world. We know that our planet is approximately 4.5 billion years old and that life is at least 3.4 billion years old. As we will show in the pages to come, we can predict that the last animals on this planet will die out as early as 500 million years from now, and possibly much earlier if another great mass extinction, similar to those of the past, once more decimates the planet. If scientists are correct that species diversity and fecundity were actually higher in the past than now, we live not in our planet's youth but in its middle to old age. Our planet is already in decline.

Such a statement must strike our egocentric species as startling, of course. It was only several hundred years ago that most thinkers assumed our planet was the center of the Universe. That Universe was a tiny place of a few nearby worlds, a relative handful of species that could fit on an ark, and a history not much older than recorded human memory. Science has had the disquieting habit of deflating our egos ever since.

The idea that we inhabit a world inferior to a past "golden age" is actually common in mythology and history. So is the idea that the world must end. While Eastern religious tradition holds that the world is unimaginably old and endlessly reincarnated, Western tradition from Egypt and Persia produced Jewish, Christian, and Islamic beliefs that the planet's history is brief and will soon come to a climactic religious cataclysm. Until relatively recently, even educated Westerners assumed our planet wasn't much older than human history and that the entire point of Creation was to produce us. Just two hundred years ago, most good Christians were taught the world was essentially as old as biblical history, or about six thousand years. Some modern fundamentalists believed its end was due at year 2000, and despite the anticlimax of Y2K, prophecies of imminent apocalypse or Armageddon persist. It is flattering to believe, of course, that our generation might live at the most critical time, the crux and end of all history. The fact that countless prophets of doom have been wrong in the past doesn't deter new ones from warning of an imminent demise on a regular basis.

Science today takes a longer view, to put it mildly—but only after a revolution in scientific thought and understanding that is still rocking world culture. At the beginning of the nineteenth century, there was no true understanding of geology, evolution, atomic theory, astronomical time or distance, relativity, or quantum mechanics. The realization that there are galaxies beyond our own Milky Way is less than a hundred years old, and the discoveries of DNA and plate tectonics are just half a century old. The discoveries of how Earth works as a system have mostly been made in the last few decades, and many in just the last few years. Using our new understanding of Earth's past to prophesize its distant future is entirely new, and entirely bold.

That is what this book intends to do, however.

It is necessary, before taking our journey into the distant future, to have some understanding of the depths of scientific time. In an accelerated age when the Vietnam War can seem like musty history to today's teenagers, grasping science time—the immense age of the Earth and Universe—is extremely difficult for our sensibilities. Nothing in our evolutionary past prepares us for such mental journeys. Yet little of this book makes sense without understanding what immense periods of time we scientists routinely fly across.

The esteemed paleontologist Stephen Jay Gould used this yardstick to get deep geologic time across to his Harvard students: if our planet's beginning is the end of your nose and its present your outstretched fingertip, then a single swipe of a file on the finger's nail wipes out all human history. Not just the history of civilization, but the entire presence of *Homo sapiens*—and indeed, the history of all our primitive hominid ancestors.

Put another way, suppose the 4.5-billion-year history of our planet is compressed into a twenty-four-hour day. On such a clock, the Cambrian Explosion of complex animal life that occurred about 530 million years ago does not show up until ten P.M., or after twenty-two of the twenty-four hours of our planet's history have already passed! Dinosaurs do not make their appearance until after eleven P.M. and are snuffed out by an asteroid or comet impact about twenty minutes before midnight. And we modern humans? Assuming that big-brained us have succeeded our hominid forebears for about one hundred thousand years, then *Homo sapiens* has been on the scene for the *last two seconds* of that twenty-four-hour day. All of recorded human civilization—some five thousand years of empires, art, politics, passion, and religion—occupies the last tenth of a second. In the epic movie of Earth's existence, our own history is but a single frame of film.

For a third analogy favored by geologists, imagine you have decided to climb Mount Everest and will compare its height to the age of our planet. You train for months or years. You trek for weeks to base camp. You ascend glacier and rock ridge, day after day, the air constantly thinner, the wind howling, the temperature dropping. Finally you assault, and gain, the summit. You've just climbed the history of Earth—and your own lifetime, in comparison to the dizzying heights below you, is represented by the height of the last snowflake at the mountain's summit.

This kind of abyssal time is as difficult to grasp as the distances of deep space. Yet scientists don't just imagine it, they inhabit it—and are increasingly confident that they don't just understand what happened to Earth over this immense history, but can predict what will happen billions of years in the distant future.

Understanding time, of course, is perhaps the earliest of all scientific quests. Ancient astronomers looked for patterns so that the sky could be used as a clock to predict planting and harvest. Time is commonly organized along astronomical lines: our year, month, and day are all based on celestial mechanics. But to recognize large units of time, we leave behind the short astronomical phenomena and look at slower-ticking metronomes: the course of organic evolution, the rise and fall of species and their dynasties such as the dinosaurs, and even the slow dance of continents across the face of the Earth. This type of timekeeping rests within the discipline of geology.

Geology is, above all, an historical science. Much of its work is empirical and, more than most other science, organizing principles are to a greater or lesser extent linked by the common theme of time. No other field of science has found it necessary to codify its own timescale of quaint and romantic Victorian origin. There is no formalized biological timescale, or chemical timescale. These and other fields of study simply use the intervals of time known to us all: seconds, minutes, hours, days, and years. Geologists, on the other hand, talk about periods and epochs, eras and zones, stages and series: the arcane subdivisions of what is known as the Geological Time Scale. If astronomers require light-years and megaparsecs to roam the vastness of space, geologists need their own language to plumb the depths of time.

The first great divisions of time used in geology come from a study of the fossil record. They are recognized and defined by mass extinction events, or global catastrophes that caused major biotic turnovers and successions. Two of these were especially dramatic. At the end of the 250-million-year-old Permian System, and at the end of a much younger, 65-million-year-old Cretaceous System, the vast majority of animal and plant fossils were replaced by radically different assemblages of fossils. Nowhere else in the layers of stone that make what is called the "stratigraphic record" was such an abrupt and all-encompassing change found. So striking were these shifts that they were used, by a pioneering English geologist named John Phillips, in 1860 to subdivide the geological timescale into three large-scale blocks of time. The Paleozoic era, or time of old

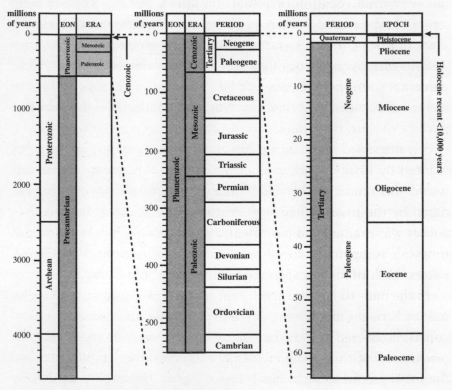

Eons, eras, periods, and epochs of geological time.

life, extends from the first appearance of life with bones or shells some 530 million years ago until it was ended by the greatest and most mysterious of all mass extinctions, the one that took place 250 million years ago. We will return to the lessons of this extinction later in this book. The Mesozoic era, or time of middle life, marks the reign of the dinosaurs and ended 65 million years ago— at least in part, scientists now believe, because of a comet or asteroid impact. This subject, too, we will return to. The Cenozoic era, or time of new life, extends from that catastrophe to the present day and is marked by the triumph of mammals. While mammals first appeared at about the same time as dinosaurs, they did not begin to dominate the planet until the Mesozoic era ended.

This system harkens back to the ornate time two centuries ago when science was the avocation of rich and bored gentry, and not yet a true vocation or profession. The timescale is filled with strange and unwieldy names, piled one upon another like an ornate stack of bricks, for in essence that is what the timescale is: words to describe the layers of rock that mark the past. One of our most fundamental scientific discoveries in history was to recognize that sedimentary rock was laid down in layers, or strata, that marked the passage of time, that similar layers of similar time could be found in widely dispersed places across the globe, and that many layers were marked by fossils peculiar to each strata: in other words, that the world was unimaginably old, and that a particular layer could be dated by the fossil remains of creatures found within it. This discovery was gradual and piecemeal, and because of that the geologic timescale is a patched-together contraption of somewhat awkward names, built of historical compromises and discovery.

In the mid- to late twentieth century a revolution in dating technology brought much more precision into this fossil timescale. Scientists discovered that the rate of radioactive decay in rocks and the orientation of their magnetized particles could be combined with the fossil record to yield much better dates. Although such dating becomes increasingly uncertain in older rocks, there is now a very sophisticated set of values for the timescale of the past 100 million years, and a fair understanding of all of the boundary ages going back to the dawn of fossil animal life some 530 million years ago. This integration of old and new has allowed Paleozoic, Mesozoic, and Cenozoic eras to be subdivided into periods. Many of these are quite familiar to us from modern culture, such as Jurassic, Cretaceous, and Cambrian. (Judging from the species of dinosaurs being cloned, *Jurassic Park* should more accurately have been called *Cretaceous Park,* by the way.) Other periods are less known, but each is defined by the first or last appearance of some form of life.

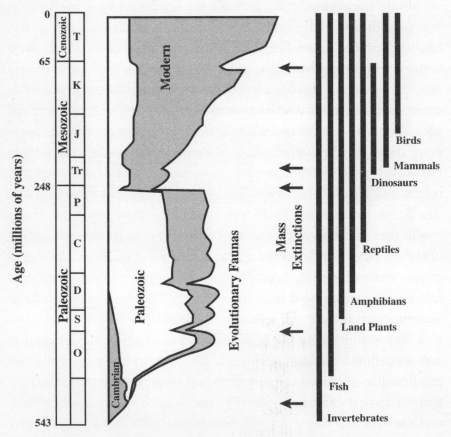

The diversity of multicellular organisms over the past 543 million years.

In addition, new eras have been defined to join the broader Cenozoic, Mesozoic, and Paleozoic eras that paleontologists first recognized. The Proterozoic extends from 530 million years ago to 2.5 billion years ago, and marks the period from the rise of complex eukaryotic cells—the kind of cells we are made of—to the first complex animals to leave a record from bones and shells. The Archean runs from that 2.5-billion-year-old point to 3.8 billion years ago, when the first primitive bacterial life, or prokaryotes, first appeared. The Hadean dates back to 4.5 billion years ago, when the Earth was formed.

We explain this timescale because, in the pages to come, we will describe hypothesized events that, when combined in sequence, will constitute the several ends of the world. These new series of eras and periods don't have names yet, of course, but they will echo the history that has come before: in fact, the "movie" of the Earth's history will seem to begin to run backward. Like the nineteenth-century pioneers of geology, we may not be able to *exactly* pinpoint when each era gives way to the next—but we have a pretty good idea about the relative order in which these various phases of planetary death will take place. Our organizing hypothesis is that past Earth history provides a model for understanding how the future might unfold, and that we are likely to, in a way, relive the past. The Earth has changed from a place that was very hot to a much cooler and water-covered world. That change was gradual. Much of the history of life is evolution to adapt to a planet where temperatures were dropping, oceans appeared, the atmosphere became filled with oxygen, and land became habitable. In the future that sequence will reverse. All scientific models predict a gradual return to a hot world where life becomes less diverse, less complicated, and less abundant through time, until the last life looks much like the first life—a single-celled bacterium, last survivor and descendant of all that came before.

Exactly when this will happen is not known, of course, but we have a pretty good idea. For fun we have proposed geologic eras for the future. Our Cenoproterozoic will be the era stretching from 500 million to 1 billion years into the future, in which first plant and animal life, and then eukaryotic cells, disappear. Our Ceno-archean era extends another 500 million years to perhaps billions of years, and is the period in which even primitive microbial life might be cooked into extinction. Our Neohadean era extends more than 5 billion years into the future to the point where the planet itself ends.

Could life make a longer run of it? There is a chance, as we shall see. Yet most evidence suggests that life has already existed on Earth longer than it will persist into the future.

THIS POWER OF PROPHECY COMES FROM A COLLABORATION OF scientific disciplines typified by the coauthors of this book. Peter Ward is a paleontologist, someone who has spent a lifetime looking at his feet to pry old bones from ancient outcrops and walk backward in time. Paleontology is a science of death, and a paleontologist but a glorified and well-educated grave robber—but to understand ancient death he must understand life. Peter's necessary marriage of geology and biology forces the kind of interdisciplinary thinking necessary for this kind of investigation.

Don Brownlee is an astronomer, another time traveler who looks up at the night sky to use that most powerful and astonishing of time machines, the light from distant stars. So immense is space that it can take thousands, millions, or even billions of years for light photons to reach our telescopes and eyes. We don't see the stars as they are today, but how they looked when light left them. Thus, when we look at the Universe we are looking into its past, and a history that suggests our future.

Don is a catcher of comets, from tiny particles that sprinkle our atmosphere (and that are called Brownlee particles in honor of their discoverer) to serving as the principle investigator for Project Stardust, a space mission launched in 1999 that will capture dust from Comet Wild-2 in 2004 and return it to Earth. Peter is the field man and scuba diver who has roamed the world, a wielder of rock drill and hammer, muddy and wet. Don is analytical and dreams of distant worlds. Peter is a synthesizer who vividly imagines the ancient world. It's an unlikely partnership, in personality and approach, and yet a productive one. We teamed up first on a book that asked the frequency of complex life in the Universe, and concluded it is rare. Now we've asked ourselves how the world will end. To do that, of course, we must understand how it began, and how it has operated up to now.

It is reasonable to ask why we set upon this quest. After all, who among us would be willing to learn the details of how—and when—our lives will end? To see the final deathbed, or fiery wreck, suicide, or drowning, or intensive care wing in a hospital? For most of us, some things are better left unknown.

We have at least three compelling reasons.

First, because we can. We're scientists. We're curious! Never before has science had such a good understanding of how our planet and Sun work and thus how they may work in the future. The temptation to project into the future is irresistible.

Second, predictions of how our planet can or will change for the worse (from the human point of view) should illustrate the perils of our present course. Our civilization is modifying the atmosphere, depleting the oceans, consuming reserves of groundwater, and contributing to species extinction. If our planet is a balanced system, how will these changes affect that system? Most people naturally assume our planet's present state as "normal," and trust that it will always go on. We hope to demonstrate that not only is our kind of planet probably rare in the Universe, but that even Earth is a place, from the perspective of science time, of potential fragility and tumultuous change. We toy with it at our peril.

Third, the bleak future we are about to present—that our planet may already be past its peak and is slowly dying—suggests that if our own species is to survive, we must either take steps to sustain

the habitability of our own world or find another at a younger stage of planetary life. In other words, we'd better know what is in store, not just on this planet but on any inhabitable planet to which we might flee. From learning how the Earth works (and how, when, and why it will cease to be a habitable world), we learn general lessons about other habitable worlds. If we can't engineer or evolve our way around our planet's inevitable decline, then we'd better go planet shopping. This book is a first, tentative step.

Projecting apocalypse, of course, is nothing new. The study of legend and literature dealing with the end of the world and Armageddon is called "eschatology," and traditionally has been left to religion and myth. Recent scientific discoveries have led to a new kind of scientific eschatology, however: contentious, exciting, and hotly debated. Space probes to the planets of our solar system, particularly Venus and Mars, have made it clear that similar rocky bodies can have far different fates that Earth. Moreover, it's clear that planets can be destroyed, for astronomers routinely observe objects in deep space that surely are the annihilated remnants of once habitable solar systems. Biology, meanwhile, tells us the physical limits that organisms can tolerate, while paleontology, the study of ancient biology, tells us that several times in Earth's history there have been "near-death experiences," when large fractions of the Earth's life suddenly died out from asteroid impact or changes in atmospheric composition and climate. Astronomy, biology, and paleontology have thus combined to deliver a sobering message: Our planet and indeed all planets have a finite span to their existence, including the life that they may bear.

WE HAVE WRITTEN THIS BOOK AS A COMPANION AND SEQUEL TO our 2000 book *Rare Earth: Why Complex Life Is Uncommon in the Universe*. While we believe that microbial life may be common in the cosmos, we argue that planets that could support complex animals and intelligent life are probably very rare. First, only a small fraction of planets would have the necessary properties allowing complex life to evolve, such as the proper temperature to retain water in a liquid state. Secondly, once in place, a planet would have only a finite time period during which conditions necessary for complex life would exist. We discussed mass extinctions, or times when the diversity of life on a planet could be suddenly reduced through short-term catastrophe such as a meteor strike or enormous volcanic eruptions or explosion of a nearby star.

Looking back, it is now clear to us that we were not yet thinking about how long habitable planets might last even if there was no sudden catastrophe. We didn't fully take into account that planets, like organisms, have life spans. Astrobiology has evolved with much new information, leading to the inescapable conclusion that habitable worlds can end through a natural evolution akin to aging. Planets have "habitability systems" that are roughly analogous to the organs of a living creature and that eventually fail like

### Earth Clock

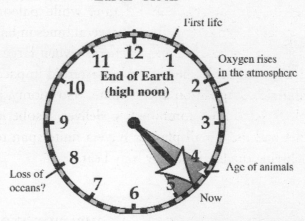

The Earth will survive for total of 12 billion years before it is either consumed or severely scorched by the Sun in its last moments as a red giant star. Earth history can conveniently be illustrated as a clock where each number represents 1 billion years. The tenure of animals and plants is remarkably short, starting at four and ending at about five.

those organs. We intend to showcase the tools and information necessary to understand this evolution, to understand where we are in the planetary cycle of aging, and to predict how life on our planet, and then the planet itself, will end.

The early and untimely death of a friend or family member always comes as a shock. We experienced one of our own when synthesizing this new research, and we used the end of her story as an analogy for this story. While our planet is a hoary 4.5 billion years old, it will exist a total of 12 billion years before being swallowed by an expanding Sun. Yet all the existing scientific models conclude complex life on Earth such as plants and animals will end billions of years before the planet itself is consumed, due to a predictable sequential breakdown of these habitability systems on our planet. As an abode for higher life, Earth is already in middle age.

# THE WONDROUS MACHINE

OUR INSTINCTUAL, SPIRITUAL, AND INTELLECTUAL REVERENCE for life has only been increased by scientific understanding of how marvelous a creation the human body is. The chemical elements used to build its tissues did not exist at the creation of the Universe but instead were forged only after the first stars formed, burned, and finally exploded. Our elemental building blocks were literally forged in these cataclysms, which were necessary to convert the simple atoms such as hydrogen and helium that appeared at the dawn of Creation to more complex ones such as carbon and iron. Billions of years later this gas and stardust coalesced to form our own Sun, planet, and eventually us. You are made of more cells than there are stars in the Milky Way galaxy, and each cell is a miniature city with its own highways, factories, waste-processing plants, and a coil of DNA recording the genetic instructions necessary to make a copy of you. The cells in turn have combined to make a complex organism that can move, eat, breathe, reproduce, repair itself, think, and dream. Yet as astonishing as our biology is, we are also mortal, forever in peril of disease, puncture, or the inevitable dissolution caused by the wear and tear of time. The heartbreaking end, of course, gives poignant joy to the beginning. We celebrate birth and youth as an antidote to death.

Our home planet is similarly complex and similarly marvelous. To understand how Earth will die we must first understand how it lives. There is strong evidence that the Universe began with an explosion, the Big Bang, about 13 billion years ago, and two-thirds of its history passed before our own solar system was formed. During this immense period, the evolution and death of stars supplemented the basic hydrogen, helium, and trace of lithium with which the Universe began with more complex atoms such as carbon, oxygen, silicon, magnesium, and nitrogen. Cast into space by winds and stellar explosions, it coalesced again and again into new stars and solar systems. Yet even today, less than 1 percent of the original hydrogen of the cosmos—the simplest atom—has been converted to more complex elements. The atoms of your body— and indeed, of our world—are products of evolutionary recycling in the cosmos.

In the beginning, our planet was the product of hydrogen gas, stardust, and gravity. Gas and dust floating in the space between stars were drawn together by gravity until they coalesced into a spinning disk that became our solar nebula, the Sun forming at the center and the planets assembling in the Frisbee-like plane of debris that circled it: called by astronomers an "accretion disk." The Earth then formed by the collision and clumping of untold rock bodies intersecting its orbital path that ranged in size from dust motes to small planets. Since astronomers have observed that accretion disks around other stars last for only a few tens of millions of years, Earth's creation must have been relatively rapid.

Earth is a "terrestrial planet," meaning that it is made almost entirely of rock and metal. The gaseous planets such as Jupiter and Saturn have no hard surfaces and were formed either mostly from hydrogen and helium or from a mix of carbon, oxygen, nitrogen, and hydrogen. They are huge, gaseous, stormy worlds. The terrestrial planets that we know of formed in the warm inner regions of the solar nebula inside what is commonly known as the "snow line." Here on Earth, temperature tends to fall as we climb in altitude, and in common usage the snow line refers to the elevation where precipitation changes from rain to snow, producing a white boundary along the side of mountains. The nebular disk also had a snow line, because it was hot in its center and cold in its outer regions. Outside of the snow line it was cold enough that water vapor, a common gas in the nebula (water is made of two of the most abundant elements, hydrogen and oxygen, and it is one of the most common molecules in the Universe), literally condensed as snow. In the inner regions of the nebula, water existed only as steam. Earth formed interior to the snow line, which meant that the bits and pieces accreting into a planet were essentially devoid of water, and also carbon and nitrogen because these materials couldn't form solids in the hot nebular gas. That meant that the critical "volatiles" so important to life were presumably carried to Earth later in its creation, both by comets that are largely "dirty snowballs" or asteroids that retain water in hydrated minerals such as serpentine. When hydrated minerals are heated they release their trapped water. Later, from the heat of the interior Earth, water could be liberated and vented from the planet's surface.

As our infant planet grew larger, it was like both a gravitational vacuum cleaner and a leaf blower. Its gravity either pulled in more and more debris to add to its bulk, or it added enough energy to passing bits of rubble to eject them toward other planets or outside the solar system. The process was a violent gestation of impact and explosion, a sweeping up of orbital litter that swept its orbital path mostly bare but left the infant Earth red-hot from the accumulated energy of countless collisions.

Late in this process of accretion planet-sized bodies as large as Mars slammed into the still-lifeless Earth. Shrapnel from such a smash-up formed a debris ring around Earth that quickly coalesced into our Moon. While our Moon is by no means the largest satellite in the solar system, it (along with Pluto's moon) is the largest relative to the mass of the planet it orbits, and its presence is one of the lucky conditions that makes our planet a rarity. Our relatively gigantic Moon has had a profound effect on Earth's history. It stabilized the tilt of the Earth's axis, allowing our planet's climate, seasons, and ocean circulation to stabilize. If our planet did not tilt, we would not have seasons as we orbit the Sun each year, and the temperature difference between equatorial regions and temperate regions would be even greater, and less hospitable, than they are now: the cycle of summer and winter helps moderate overall temperature extremes. Yet if we tilted too much, or the tilt shifted, our climate would become more extreme or be thrown into chaos. This may have happened to Mars, in fact, allowing that planet to lose its oceans. The Moon keeps Earth's tilt at a stable angle, providing needed stability for complex life.

Moreover, by helping to generate large ocean tides, the Moon created the alternately wet-and-dry tidal zone that encouraged the transfer of life from the sea to the land. This tidal pull has also slowed the frenetic early spin of the Earth, lengthening our planet's day to twenty-four hours. This dance of two worlds has not just braked the spin of the Earth, it has pushed the Moon farther away, and these twin trends continue today. Just over the past 500 million years, for example, the length of Earth's day has increased by about 10 percent because of the Moon.

The Moon has also left a record of its violent past that has been erased by erosion from the surface of the Earth, but that gives us invaluable information about the early history of our own planet. The "Man in the Moon" was created by violent impacts and later magma flows named mare (Latin for sea) that created the pattern of light and dark that we call a face. The "man's" left eye is the basalt-filled Mare Imbrium, a 3.9-billion year-old impact crater formed by an incoming projectile that was a hundred kilometers wide. The

Moon's Atkin Basin at our satellite's South Pole is a crater that is more than 2,400 kilometers in diameter. By going to the Moon, astronauts were able to collect geologic evidence that has told us a great deal about the early history of the Earth. The Moon rocks serve as another kind of time capsule.

The history of our planet is little known for its first 600 million years because there are no rocks of that age left on its surface. Discovery of tiny grains of the mineral zircon has provided some insight into the early Earth, even though the rocks that carry them have long been altered beyond the stage where they can be dated. Using the decay products of the radioactive thorium and uranium, the zircon ages can be determined to accuracy of about 1 percent. The analysis of these rare grains, miraculous survivors of the Earth's earliest history, provide evidence for oceans and continents as early as 4.2 billion years ago. Still, there's a difference between inferring from a grain of zircon and studying an entire satellite. The Moon, with no air or oceans to erode its surface, provides a surrogate record of the early violent history of Earth.

Even as the face of the Man on the Moon and other giant craters were being formed, the same kind of stupendous collisions were occurring on Earth. Each great impact was capable of repeatedly blowing away part of our planet's fragile atmosphere and vaporizing newly formed oceans into steam. This cataclysmic infancy would seem to spell a bleak future for any life and yet this appears to be the normal birthing process for habitable planets. A more hostile environment can scarcely be imagined. In fact, however, these smash-ups appear to have contributed to the conditions for life to later arise. Had our planet formed in a more gentle way, with only the gathering of small rocks in its orbital zone, it would have formed cold and dead, because the energy of the falling bodies would have been small enough to radiate back into space. Instead, Earth was hit by monstrous, planetary-sized chunks of rock and ice that cratered so deeply into the interior that their heat, water, and future atmospheric gases were retained. As Earth reached its final size, the great power delivered by these bodies melted the Earth's surface down to depths of hundreds of kilometers, creating an ocean of magma. This process also occurred on the Moon and produced the ancient bright regions that can easily be seen today simply with a pair of binoculars. The bright lunar material is dominated by crystals of feldspar, a mineral that formed in the ancient lunar magma ocean, that floated to the surface forming the still-preserved crust of the moon. In Earth's earliest history, its surface was molten rock and its atmosphere was a torrid mix of steam and other gases. It was truly a hellish place, although when the impact rate diminished sufficiently the surface cooled, leaving a rock crust covered by warm bodies of water. For hundreds of millions of years occasional giant impactors continued to hit the planet. The embryonic seas were repeatedly annihilated—with each impact of a body larger than a hundred kilometers in diameter—flashing them back into vapor; they recondensed again and again, seas forming within only a few thousand years of a great collision.

Did life form repeatedly in these early oceans, only to be destroyed by the next impact? Perhaps, but how would anyone know? But finally, after 700 million years—a stormy eon longer than there have been complex animals on Earth—the large collisions were mostly over. While Earth still gains 80 million pounds of rock and metal from interplanetary space each year, today's meteors are nothing like the bombardment that occurred so long ago. Much of our annual added weight today is mere dust. The main period of giant, ocean-boiling impacts was finished 3.9 billion years ago, leaving our planet as a reasonably safe and relatively stable platform for life for billions of years to come. But not too stable: the heat of our planet's interior ensured not only that Earth would have an atmosphere and oceans, it began the plate tectonics that meant its two-mile-deep lifeless ocean would be broken by dry land. Heat and pressure began creating rock lighter than the ocean floor that would ultimately emerge as floating, accreting continents.

This was vital because land may be critical for higher intelligence to evolve. Whales and dolphins have large brains and some level of sophisticated communication, but they give little evidence of the complexity of our own species. Indeed, it has been argued that human-level intelligence and technology would never develop on a purely water-covered planet. There would be neither need nor opportunity. Yet it is likely that the early Earth was entirely covered with water, and many Earth-like planets in the Universe are probably water-covered. The average depth of the ocean is about 3 kilometers, and without geologic forces, there would be no irregular topography to rise above the surface. Large volcanoes such as those that formed the Hawaiian Islands are probably not sufficient for evolution to take place, because they are relatively small and last only a few million years before weathering and gravity pull them below sea level. What life needed to advance was a long-lasting continent, and Earth's internal heat energy created the conditions for granite and andesite to form and "float" above the heavier, denser basalt of the ocean bottom. Alternately, continents may have started when giant collisions generated large magma bodies and mountain ranges of ejected debris. In any event, the early Earth had few if any continents; they grew as geologic processes continued. We will go into the full importance of the plate tectonics that build continents later in this book, but for now we will simply note that today's continental surface area is twice as big as it was 2 billion years ago, and it will continue to slowly grow into the future.

Earth's mix of land and ocean is unique in the solar system and it is a vital property that has allowed the Earth to be relatively stable and habitable for as long as it has. The combination of water and land allows carbon dioxide cycling to serve as a long-duration thermostat moderating temperature swings that would otherwise be caused by the composition of the atmosphere, changes in the Sun, and other factors. Surface water and land that rises above it is probably the most important asset that a planet can have if it is to ever harbor advanced life.

Land was one essential for life to evolve. Air is another. Without an atmosphere there would be no life on Earth, and without sufficient oxygen there would be no higher life. Our atmosphere's composition over Earth history is one of the reasons why our planet has remained a life-supporting habitat for so long. Today, the atmosphere is highly controlled by biological processes and it differs greatly from other terrestrial planet atmospheres. Mercury has essentially no atmosphere at all. Venus has a carbon dioxide atmosphere a hundred times denser than Earth's. Mars has a carbon dioxide atmosphere a hundred times less dense. Earth is intermediate, comfortably intermediate. Earth's atmosphere was initially vented from its hot interior, but it has changed a great deal over time. Even viewed from a great distance, the Earth's unusual atmospheric composition would provide a strong clue for the presence of life. Composed of nitrogen, oxygen, water vapor, and carbon dioxide (in descending order of abundance), it is not an atmosphere that could be maintained by chemistry alone. Without life to replenish it, the free oxygen would rapidly diminish, because some would be consumed oxidizing surface materials and the remainder would react with nitrogen, ultimately to form nitric acid. To an alien astronomer, the Earth's atmosphere would provide very strong evidence of a vigorous ecosystem capable of controlling the chemical composition of its atmosphere.

WERE WE TO GO BACK TO THIS INFANT WORLD SOME 3.9 BILLION years ago, it would still seem an unlikely cradle for the future diversity of life. There was essentially no oxygen in the atmosphere, meaning we couldn't breathe. Without oxygen there was also no formation of protective ozone high in the atmosphere, leaving what little land had emerged from the waves, and water itself, open to merciless bombardment from the Sun's ultraviolet radiation. To make things even worse the young Sun emitted even more UV than it does at present. While the young Sun's total energy output was 30 percent less than it is now, the Earth's surface temperature was probably hotter because the dominant atmospheric gas was probably carbon dioxide, the famed "greenhouse gas" we are now ejecting into the atmosphere by burning gas and coal. This trapped the infant Sun's heat and raised the temperature. Above this hot, breathless, ocean-dominated world, the Moon would look gigantic, since it was then close to Earth and has slowly spiraled outward ever since. The only life, if it existed at all, was microbes. Alien astronauts, had they visited, could be excused for writing our planet off.

Yet despite how bizarre our planet would have looked in its infancy, two criteria crucial to future life were already present.

First, the Earth was an ideal distance from the Sun and it could retain its oceans of water. This was critical. Water, a simple molecule of one hydrogen and two oxygen atoms, is important because it remains in liquid form in the same temperature range at which life can exist. This is actually unusual among the liquid chemicals commonly found on planets or moons. Ammonia, methane, ethane, carbon dioxide, sulfur dioxide, and nitrogen exist as liquids only at far lower temperatures. For example, Saturn's moon Titan may have oceans of methane, but no life is likely to exist there because the temperature is minus 150 degrees Celsius, or far colder than Earth's lifeless South Pole. Only water stays useful by being a liquid medium in the same temperature range that works best for biological chemistry.

Water is also a remarkable solvent that erodes to create soil, temper atmosphere, and works as a happy medium for the chemical reactions necessary to sustain life. As a result, most organisms are essentially just complex bags of water: we humans are 70 percent $H_2O$ and our fluids carry a saline memory of the sea inside us.

Finally, frozen water floats on top of liquid. This is important because if ice sank it would settle to the bottoms of oceans and ponds where it would never melt, and eventually all the Earth's water would freeze solid. Instead, frozen water insulates the liquid water underneath, keeping it at a habitable temperature.

(Our oceans are 3 percent salt, and the reason for this has long been debated. Was salt washed into the sea from the land? Or were the oceans salty virtually from their beginning? While it is commonly believed that the salt content of the oceans has increased in time, Paul Knauth of Arizona State University has pointed out that the opposite is actually the case. The oceans have always been salty and in time more and more salt has been removed to deposits on land, in the form of salt domes and similar geologic deposits.)

The second criteria for life was the brisk heat of the planet's interior. It seems odd to say that air and water come from rocks, but this is essentially true: the water and atmospheric gases contained in the chunks of debris that hurtled in from space are released by our planet's constant volcanism to form the thin envelope of gas and water that life inhabits.

The Earth's proportion of water to its weight is small, as we would expect from a rocky planet inside the snow line: it is less than one-tenth of 1 percent. Where does even this modest water supply come from? Outer space, specifically regions beyond the orbit of Mars. Carrier bodies include asteroidal and cometary materials as well as a few objects of lunar to Martian size. Some asteroidal meteorites that collided with the infant planet were more than 10 percent water, in the form of hydrated minerals such as serpentine. The comets that hit were dirty snowballs, perhaps half water and half rock and carbon. Some of the meteorites that formed in the asteroid belt contain up to 6 percent carbon and up to 20 percent water of hydration. All brought water to Earth and locked it beneath the surface until vented by volcanoes. And even as fresh magma is spewing in one spot, at another point old crust is being sucked back into the hot interior and remelted. This constant recycling of Earth's crust has been absolutely vital in balancing the composition of our atmosphere and sustaining life, as we shall see.

We do not know exactly when, how, or why life first arose on Earth. Perhaps it emerged at deep-sea vents, a mile or more below the surface. Perhaps it generated in shallow pools. Perhaps it arrived from space via meteorite from an older, more biologically advanced Mars: scientists continue to debate whether a Martian meteorite found in Antarctica bears fossil evidence of microbial life. In any event, traces in rock show that primitive bacteria were present on Earth at least 3.6 billion years ago, or almost as soon as conditions allowed once the catastrophic, planet-sterilizing bombardment from space ceased. Evolution had begun.

There is little doubt about what the earliest life on Earth would have looked like: it would have resembled modern-day bacteria in size and appearance. When we do find the oldest fossils of this early life, we see entombed in rock cells that could belong to any number of living microbes. The common shapes of bacteria—rods, balls, and spirals—evolved early, and stood the test of time. What *has* changed is not the shapes of these cells, but their inner workings—mainly, how they use energy. Living cells are biochemical factories that are surrounded by porous membranes and get energy either directly from sunlight or indirectly by eating other organisms or floating compounds those organisms have produced. Cells have become steadily more sophisticated in their ability to harness sunlight or ingest and transform other molecules to power themselves.

These early bacteria were prokaryotic cells, and their descendants thrive on Earth today. A prokaryotic cell is much more primitive than the cells that make up plants and animals, protozoa or fungi. They are mostly small, about 0.2 to 10 micrometers—far smaller than most cells with a nucleus. If you lined up five hundred typical bacteria in a row, they would stretch no farther than the thickness of a dime. They take their name from the Greek words for "before kernel," or nucleus. Unlike the cells that make up our bodies, prokaryotes have a single strand of DNA that is not bound by any membrane—it is thus free in the cell—and have no internal organelles, such as the mitochondria that typify animals and plants.

They have no complex membranes in the cell, and for locomotion have, at most, simple flagellum. In comparison to the complexity of a modern cell they are like a rowboat to an ocean liner, and yet they show an enormous range of ways and environments in which they can survive. Bacteria can be found in places as diverse as hot springs and deep underground rock, and they dominated the Earth alone far longer than dinosaurs and mammals combined. Averaged over the full lifetime of Earth, they will be the dominant form of life.

By 3 billion years ago bacterial colonies of prokaryotic cells began coalescing into matlike communities called stromatolites, primitive and curious structures that we will meet again in the distant future. The top layer of a stromatolite is made of a photosynthesizing bacteria called cyanobacteria, or blue-green algae. (This should not be confused with other kinds of algae, such as seaweed, which are in separate biological kingdoms.) Like the land plants that are still billions of years in the future, this primitive bacteria could harness sunlight and water, breaking water's chemical bonds and releasing free oxygen into the air and water. They were simple, successful, and dominated early shallow bays and seas. And with that success came a revolution: life was creating the conditions for its own evolutionary advance. Oxygen, although corrosive and poisonous to the creatures excreting it, was like an injection of speed that would allow a quicker and more complex biological chemistry. Unfortunately, the presence of oxygen would allow the evolution of more advanced creatures that would ultimately eat the stromatolites. So here is a lesson from an early global polluter: anaerobic bacteria was giving off a poisonous gas (oxygen) that would spell its own doom.

Stromatolites can still be found today, but only in nonmainstream environments such as very salty swamps or very hot lakes. The most famous occur in a series of salt pools along the Sinai shore and at supersalty Shark Bay on the western coast of Australia. A combination of restricted water flow and high evaporation from the tropical sun has caused this bay to be up to twice as salty as the normal ocean, inhibiting the snails, chitons, and sea urchins that would normally eat the stromatolites. We can thus see, touch, and ponder some of the earliest inhabitants of Earth, bacterial mats that are far older than dinosaurs. These inert lumps helped change the world.

Do not feel too sorry for the stromatolites, however. As we shall see in the pages ahead, their time will come again.

The oxygen they produced from photosynthesis changed the chemistry of the oceans and land. The transformation took a long time. Initially, oxygen reacted with iron on both land and at sea, the oxygen being consumed and the iron in seawater being lost, and it wasn't until 2.2 billion years ago that a new equilibrium was reached. Once that happened, oxygen could begin to accumulate in the atmosphere, allowing creation of a layer of stratospheric ozone that protected any future life—especially life on land—from ultraviolet radiation. The new abundance of oxygen also made unintended room for higher life. Oxygen's usefulness abetted the evolution of the eukaryotic cell, where all the genetic material is encased in a nucleus within the cell. Your body consists of 50 trillion to 100 trillion eukaryotic cells working in unison.

The oldest known fossils of an organism that appears to have achieved the eukaryotic grade of organization have been found in banded-iron deposits located in Michigan. The fossils themselves are about 1 millimeter in diameter, and are found in chains as much as 90 millimeters long. This is far too large to be either a single-celled prokaryote, or even a single-celled eukaryote. This creature, named *Grypania,* is preserved as coiled films of carbon on smooth sedimentary rock. These early eukaryotes may have been rare, for other eukaryotes do not appear in the fossil record for 500 million years after this first appearance.

For the period between 2 billion and 1 billion years ago, there are few notable achievements of life to be found as fossils in the rocks. Like today's amoebae and paramecium, the lack of skeletons has rendered most life from this time invisible in the fossil record. The first common appearance of eukaryotes begins about 1.6 billion years ago when microscopic fossils called *Acritarchs* begin to appear in the geological record. These are spherical fossils with relatively thick organic cell walls. They are interpreted to be the remains of plankton algae, which used photosynthesis to respire and lived in the shallow waters of the world's oceans. Presumably their single-celled predators evolved as well. Whole armadas of these floating pastures and their grazers lived and died in this seemingly endless epoch of geologic time. The open ocean would have had little life, but the coastal regions rich in nutrients would have been alive with microscopic life. At 1 billion years ago the tempo of evolution increased, and the first red and green algae—ancestors to modern seaweed—began to show up.

The evolutionary steps required to move from single-celled plants and grazers to complex marine animals are numerous. Respiration, feeding, reproduction, waste removal, information reception, and locomotion all require the integration of many cells working together. Cells had to divest themselves of their tough outer coating and, naked, adhere to one another so they could exchange living material and information. Cells began to unite for mutual benefit.

This spelled bad news for the stromatolites, which by 600 million years ago were nearly absent from Earth. They were literally being eaten out of existence, for a great biological revolution was taking place, one creating entire suites of organisms adapted for utilizing plants as food. These newly evolved grazing animals (many looking like small worms) used the stromatolites as dinner. About 700 million years ago a steep decline in stromatolite diversity began to take place. Newly evolved plant eaters were surely its cause, although these grazing creatures also left no fossil record of their existence—they were too small, and had no mineralized skeletons that could fossilize.

How did one emerging multicelled plant and animal lead to

another? We don't know in any detail. The fossil record for these early soft-bodied experiments is scant, and the evolutionary sequence remains unclear; what is clear, however, is that about 530 million years ago there was a sudden proliferation of marine animals experimenting with a vast variety of body plans. So startling is the appearance of these animals with shell-like carapaces that paleontologists call this sudden proliferation of fossils the "Cambrian Explosion." The shallow seas were suddenly full of crawling, squirming, swimming, burrowing creatures.

Many of these early experiments with various body architectures quickly died out, but the successful ones continued to evolve and spread into new ecological niches. Some 400 million years ago, the first plants and animals emerged onto land. Called the Silurian period, this was a time when life radiated from the oceans to swamps, then coastal lowlands, and then beyond. Again, the adaptations needed to leave the sea were manyfold. Plants had to develop an outer cuticle to avoid drying in the sun. They had to develop roots to mine water and minerals. They had to develop stomata, or openings in the cuticle, through which to breathe. As they rose taller in competition with one another, they had to develop vascular tissue to transport water and nutrients from the ground to their tops. They had to develop seeds (eggs) that would not dry out until fertilized and growing.

Animals followed that had much the same challenge: resisting desiccation, breathing, moving, and so on. They had to grow legs, develop lungs, and improve their senses. The challenges were equaled by benefits, however. Plants would exude oxygen beneficial to animals, and animals would exhale carbon dioxide necessary for plants. Each consumed the other, in an endless cycle of eating, death, rot, and rebirth. Life essentially conspired to help itself, in the process making Earth ever more bountiful.

This familiar and heartening tale of advancing life will be returned to in more detail later in this book, because we believe that this story of rising sophistication and complexity is going to reverse as our planet ages. Instead of becoming evermore diverse and intricate, life on Earth will become more primitive as conditions for it decline. In fact, we believe our planet may already be in decline. This disturbing idea is very, very recent.

THE HISTORY OF BIODIVERSITY—OR THE VARIETY, RICHNESS, and complexity of the Earth's plants and animals—was first considered in the work of John Phillips, the pioneering geologist who first divided the geological timescale into the Paleozoic, Mesozoic, and Cenozoic eras. Phillips, who published his monumental work in 1860, recognized that major mass extinctions in the past could be used to subdivide geological time. But Phillips did far more than recognize the important of past die-offs and name past eras: he proposed that diversity in the past was far lower than in the modern day, and that the history of biodiversity has been one of an increase in the number of species, except during and immediately after the massive extinctions. He proposed that mass extinctions slowed down diversity, but only temporarily. The implication was that the world is getting steadily richer in species.

This was a novel and important idea, but a century passed before the topic was again given scientific attention. The ensuing work since then gives a good idea of how science works, always challenging and amending its ideas.

In the late 1960s, paleontologists Norman Newell and James Valentine again considered the problem of exactly when, and at what rate, the world became populated with species of animals and plants. Both wondered if the *real* pattern of diversification was of a rapid increase in species during the so-called Cambrian Explosion of about 540 million to 520 million years ago, followed by an approximate steady state since then. They argued that while the variety of fossils indeed seemed sparser in the past, as Phillips had recognized, perhaps this was only because older fossils are harder to find, given the age and erosion of rocks. It was sampling bias, not rising biodiversity, that caused the rising diversification seen by Phillips.

As a crude analogy, suppose all humans were snuffed out tomorrow and an alien archaeologist came to Earth and tried to determine which civilization used more cups—the ancient Romans or the modern Italians? He would find more modern Italian pottery, of course, but only because it is of more recent vintage and hasn't been broken or lost. In a similar vein, perhaps we find fewer fossil species in the distant past because the restless surface of the Earth has obliterated them over time.

This view was forcefully argued by paleontologist David Raup who, in a series of papers, pointed out that older rocks experience more alteration through recrystallization, burial, and metamorphism. There are fewer older rocks, and thus fewer older fossils—but that doesn't mean there were really fewer species, Raup contended.

The University of Chicago's Jack Sepkoski sought to resolve this debate in the 1970s by compiling a massive data set on fossils with the help of colleagues and students. He looked at marine invertebrates, or sea animals without backbones, as well as land plants and vertebrate animals. His numbers seemed to vindicate Phillips's early view. In particular the curves discovered by Sepkoski showed quite a striking record, with three main pulses of diversification. The first was Cambrian fauna, the second was marine invertebrates of the Ordovician era, and a third was a sharp rise in more recent time to produce the high levels of diversity seen in the world today.

While the Sepkoski work seemed to show a world where runaway diversification is a hallmark of the late Mesozoic (or dinosaur) era into the modern day, worries about the very real sampling biases described by earlier workers persisted. Accordingly, paleontologist Richard Bambach and colleagues decided to test the diversity patterns by looking at the number of fossils within a particular environment. They did this by examining counts of species found within individual outcrops of fossil rock. For example, his team looked at the record of outcrops that were deposited in shallow marine settings, with each individual outcrop presenting a single data point. They then posed the question: Does the diversity through time increase in similar environments? This test seemed to show that while some environments (such as the stressful surf habit of modern and ancient shores) showed no change in diversity through time, offshore shelves revealed a diversity pattern that seemed to support the hypotheses of Phillips, and later Sepkoski: that diversity increased markedly in the late Mesozoic, and continues to do so to the present day. The number of species on Earth has generally been rising.

Science never sleeps, however. Computers quicken, and old data can be reexamined. As the twenty-first century dawned, the issue was reviewed with a more comprehensive database; to the profound surprise of many, a quite different pattern emerged. We may have reached a steady state of diversity some time ago! Scientists also learned that biological productivity—the total amount of living plant and animal tissue on Earth—appears to have been higher 200 million to 300 million years ago, when the planet was warmer and richer in atmospheric carbon dioxide than it is today. (We will return to this finding.) This is a seminal discovery. It would mean that irrespective of the number of varieties of life, there would be *more* life in the past—a great volume of living tissue on the planet.

It may be, then, that diversity peaked early in the history of animals and, in contrast to all views since the time of Phillips, has remained in an approximate steady state since. While the colonization of land led to many new species, the proliferation ended by the late Paleozoic, perhaps 300 million to 250 million years ago. Since then the number of species on the planet has been approximately constant, or perhaps even dropping. The planet may have been a richer place at the time of the dinosaurs, when carbon dioxide levels that plants require were higher and temperatures were generally warmer, but not too warm: a tropical world that may have exhibited the lushness of today's tropics. If true, the implication of this for our thesis is important: perhaps our planet, rather than still growing in biodiversity totals, has already peaked.

What about the recent but staggeringly important role of humans in dictating levels of biodiversity? These models have been constructed looking at the deep past, and theorize a world where evolution runs according to the "old," i.e., prehuman rules. But it is naive to believe the processes acting in the long period of pre-humanity will take place in the same fashion in our world where humanity has dominion over so much happening on the planet.

What is the future of biodiversity on the future Earth? Here the presence of humanity clouds our crystal ball, and astrobiological models are of little use. Humans are a wild card. We promote bio-diversity in some areas and curtail it in others. The only certainty is that the biota making up our world will be different. Even in the near future the makeup of species and their distribution, rela-tive numbers, and relationships between one another will have changed, and by the far future the accumulated changes may be breathtaking, for there can be no doubt that the evolution-ary forces—perhaps highly affected by humans—will create new species and varieties, resulting in a global biotic inventory of species on Earth different from that of today. However, in this Age of Humans, the old rules resulting in the biodiversity of the current world are unquestionably changed by the presence of humanity.

It is an unambiguous fact that very early on our species learned to manipulate the forces of evolution to suit its own purposes, creating varieties of animals and plants that would never have appeared on Earth in the absence of our will. Large-scale bioengi-neering was under way well before the invention of written lan-guage. We call this process "domestication," but it was nothing less than efficient and ruthless bioengineering of food stocks—and the elimination of those species posing a threat to the food stock. Once the new breeds of domestic animals and plants became necessary for our species' survival, wholesale efforts toward the eradication of the predators of these new animals were undertaken.

The modern efforts at biological engineering are but an extension of the earlier efforts of domestication. Until the end of the twentieth century the natural world had never evolved a square tomato, or any of the numerous other genetically altered plants and even animals now quite common in agricultural fields and scientific laboratories. Just as physicists are bringing previously unseen elements into existence in the natural world through technological processes, so too has our species invented new ways of bringing forth varieties of plants and animals that would never have graced the planet but for the hand of man. The new genes created and spliced into existing organisms to create new varieties of life will have a very long half-life; some may exist until life is ultimately snuffed out by an expanding Sun some billions of years in the future.

Humans have thus profoundly altered the biotic makeup of the Earth. We have done it in ways both subtle and blunt. We have set fire to entire continents, resulting in the presence of fire-resistant plants in landscapes where such species existed only in small numbers prior to the arrival or evolution of fire-branding humans. We have wiped out entire species and decimated countless more either to suit our needs for food or security or simply as an accidental by-product of changing the landscape to favor our new agricultural endeavors. We have changed the role of natural selection by favoring some species that could never otherwise survive in a cruel Darwinian world over others of estimably greater fitness. We have created new types of organisms first with animal and plant husbandry, and later with sophisticated manipulation and splicing of the genetic codes of various organisms of interest to us. The presence of humanity has begun a radical revision of the diversity of life on Earth—the number of species present, and their abundance relative to one another. We have created not only new ways of producing animals and plants through brutal *unnatural* selection, but we have also manipulated the most potent force of evolutionary change—the phenomenon of mass extinction. Humanity has even created a new mass extinction that is different from any that has ever affected the planet.

Extinction is the ultimate fate of every species; just as an individual is born, lives out a time on Earth, and then dies, so too does a species come into existence through a speciation process, exist for a given number of years (usually counted in the millions), and then eventually becomes extinct. The fossil record has tabulated random extinctions taking place throughout time. But the rate at which these "random" extinctions have taken place through geologic time turns out to be remarkably low. In order to account for this, Chicago paleontologist David Raup introduced the term "background extinction rates." Raup has calculated that the background extinction rate during the past 500 million years has been about one species every four to five years. In contrast, conservationist Norman Myers of Oxford has estimated that four species *per day* have been going extinct in Brazil alone for the past thirty-five years. Biologist Paul Ehrlich has suggested that by the end of the twentieth century, the extinction rates were measurable in species per hour. These rates of extinction exceed anything known in the deep past. If continued for an appreciable period they will reduce the world's biodiversity to levels under those of anytime during the past 100 million or even 200 million years, and will certainly contribute to the planetary reduction in biodiversity that will ultimately occur through natural affects of heating and carbon dioxide reduction—but millions of years sooner than would occur if humans were not on the planet.

CERTAINLY WE HAVE SHOWN THAT THE CONDITIONS FOR THE Earth's rich diversity did not pop into being, but instead were the result of billions of years of geologic processes, the modification of the atmosphere, and evolution. A complex system has arisen in which plants and animals provide services to one another and in which photosynthesis and animal respiration have kept oxygen levels roughly constant at 21 percent of our atmosphere for the past 400 million years. Life, in a sense, has terraformed the planet, and it is our hope and belief that we inhabit a young and vigorous world where life's greatest moments are still ahead. But what if the earth system that sustains this planetary perfection begins to break down? What if the Earth is already well into old age? This is something that the planetary "doctors" are beginning to believe. What is the prognosis from the Earth's last "checkup"?

A trip to the doctor's office means—and involves—different things depending on our stage of life. As babies we are in for observation and inoculations. At this stage no doctor is overly worried about the squalling baby being too fat. But we are concerned about those babies who are not fat enough. Predictions are usually made during such visits, predictions that are based on a series of observations through time: the child will reach such and such height or weight at the cessation of growth—predictions usually about the future success of the organism. Later in childhood, trips to the doctor take place only in reaction to events or changes: mainly childhood injuries or the occasional severe flu. Even as young adults we rarely see the insides of a doctor's office unless the uncommon accidental cancer or life-threatening injury intrudes. But at some indefinable age—that strange ontogenetic stage known as "middle age"—visits to our doctor change in nature and importance. For the first time the emphasis switches to a new and more ominous goal—predicting when and how your life will end, and how to delay that inevitability. Now the doctor's ministrations deal with measuring specific attributes of organ system function as a means of predicting when they might fail. Those of us in this time know these travails all too well: blood pressure, cholesterol, kidney function, and liver function. The search for cancers. The workings of the heart. The degradation of the reproductive system. The changing chemistry of the body and its hormones. When examined year after year, these readings provide the doctor with a rough idea of how the particular body being examined is declining toward death—for *that* is the central fact of life in middle age, that a slow decline has set in, with an invariable end. The only question is how long it will take, which system will fail first, and what variables will either extend or subtract from the time yet allotted.

In a rough sense this analogy extends to planet Earth as well. What we might call the "Age of Animals" is well into middle age. It began, as we have seen, only some 500 million to 600 million years ago—on a planet 4.6 billion years in age. There are specific tests and measures that astrobiologists—the doctors of planets—can take (and have been taking) that, when averaged over time, yield important clues about how much time is left not only for animal life but for all life on Earth, and even for the Earth itself. And in some ways it is easier to predict the ends of the Earth than the end of specific organ systems of a human, for the Earth has fewer major systems that control the fate of life on the planet. Ultimately there are but two essential measures that will control the fate of life on the planet: surface temperature and the amount of carbon dioxide in the atmosphere. And like slightly mad doctors prodding a particularly interesting patient, the planetary doctors have been measuring vital signs in the present day—and comparing these data to their equivalents stored in the Earth's rock record. The result is almost medical: there is a good record of the temperature of patient Earth through time, and of its vital chemistries. Predictions of how long this patient has, and what will end its various components, can indeed be made—and have been made. The results are surprising—but such is always the case when a diagnosis of approaching death is made.

The Earth, as a habitat for animal life, is in old age and has a fatal illness. Several, in fact. It would be happening whether humans had ever evolved or not. But our presence is like the effect of an old-age patient who smokes many packs of cigarettes per day—and we humans are the cigarettes.

· 3 ·

# THE LIFE SPAN OF
# HABITABLE PLANETS

THE SMALL WOMAN AWOKE ALONE IN THE DARKENED ROOM, eased from fitful sleep by the insistent beat of predawn rain on her shuttered windows. Ruth Ward was well past eighty, irrevocably in the time of life that we call old age. The raven black hair of her youth had long ago turned white, and her once erect carriage was bent with age. In her twenties she had been beautiful in a way quite unique: of whitest skin and blackest hair. Time had changed all of that, however. Her body, which had faithfully served a well-lived life, was breaking down, and the pain of this darkened February morning was different from the familiar arthritis and degenerating circulation. Her youngest daughter, coming to see her as she did each morning, found her mother cold, pale, and in obvious distress. Although Ruth protested feebly, she was bundled into a car and taken to the emergency room of the nearest hospital.

The medical team found a woman sick, dehydrated, and show-ing evidence of increasing systemic failure. Radiographs revealed the presence of a ruptured intestine, causing food and bile to leak into her gut cavity and starting a potentially life-ending infection. Her immune system was being battered in a cruel war—human against microbe—and was losing the fight.

The surgery took three hours. Tubes were inserted for the addition of fluid, the removal of bile and urine, and the administration of pain medicine. She was wired for cardiac output, respiratory rate, and blood pressure. An air tube was inserted in her trachea, and she was put on a respirator to help her lungs. Now part human and part machine, she was wheeled after surgery into a critical care unit, surrounded by her worried children. Her battle was only beginning, and each victory seemed to trigger a new defeat.

Great volumes of antibiotic-laced liquid had to be pumped into Ruth's body to help the immune system suppress the infection. But while aiding one body system, these fluids overloaded her excretory system. Already stressed by the long surgery and anesthesia, her kidneys labored to keep up.

Early the next morning, her vital signs remained positive. Her kidney function was now improving (though was still a source of concern) and her heart rate was strong. But as the morning wore on, the good news was overshadowed by the misery of the woman in the high-tech bed. Ruth fitfully awoke to find a plastic tube down her throat. Her nervous system was receiving a constant gag reflex, signaling to her other body systems that the entire organism was in extreme danger. Hormone systems pumped out adrenaline and other fight-or-flight chemicals that nature had evolved for quite different life-threatening situations, sapping her strength and energy reserves.

By the second night her kidneys continued to improve but her body's overall strengths and energy reserves were increasingly depleted. Her metabolic system was stressed by a lack of energy and food. Her lymph system was struggling with infections, and antibiotics were suffusing into her partially collapsed lungs. The biological crisis was jumping from one interlocking system to another. As her respiratory system struggled to keep up, her heart began to labor. Doctors registered more concern than optimism for the first time.

By early the next morning, another ominous sign appeared. One of the routine blood tests detected an increase in white blood cells, a telltale sign of infection. The fight to breathe was taking energy from other vital systems, and digestive system bacteria were rapidly growing in numbers. Her blood and lymph systems swung into action, fighting the resurgence of infection in her peritoneal wall. Doctors tried to increase the amount of antibiotics in her body, but they did so with caution. Fixing this new danger would cause stress on other systems that were already laboring. The fight for Ruth Ward's life was now a system-by-system struggle, some improving, some deteriorating, all interconnected. As energy waned, one by one the various systems began to fail, slowly but inexorably. Cells and tissues still lived, but the most complex biological assemblages, the organ systems, had begun to shut down.

Early in the predawn darkness of a long Seattle winter night, she lay quietly, breathing gently. A nurse looked at the monitors, noticing with concern the slow heart rate. She comforted the tired family member then doing watch. The two of them shared a quiet word, then looked again at the monitor.

Ruth Ward's heart had finally stopped. Surging blood slowed and stilled. Cells were denied oxygen. Electrical signals flickered out. As hours passed the many varieties of death overtook her body as very specific chemical changes began to sweep through the body. Cell walls, organelles, and nuclei began to break down. Amino acids were transformed. Armies of bacteria began to rise and succeed one another. The elements of Ruth Ward, once forged in stars, began a new cycle of transformation. This recycling would be cut short two days later with cremation, a flash of heat converting those elements from an organic to an inorganic formation.

Ruth's birth, growth, life, and death are analogous to our own planet. As we saw in the last chapter, Earth did not spring into being in its present state; instead, it took billions of years to develop into what it is today. Similarly, it will not sustain today's environment forever. Our planet will break down as the human body does, in a complex series of processes that can be forecast as surely as our own deaths. Ruth died as we believe habitable planets die: slowly, over time, with the more complex parts going first, leaving behind ever-simpler factions of life.

The end of Mother Earth may not be so different. First higher life is removed, ecosystem by ecosystem, as our planet dies. Left are ever-simpler microbial communities. Eventually even these will be seared out of existence by the crematorium that will be the inner solar system of an aging and expanding Sun.

Is the history of Earth analogous to that of the human body? Is the Earth "alive" in the way an organism is alive?

We certainly know that Earth has changed through time. If we could watch pictures of our planet at 500-million-year intervals, we would see a gradual change from a hot, molten world to one with oceans and atmosphere. Life, as we have seen—and will see in more detail as this book progresses—was part of this transformation. So is Earth similar to Ruth Ward, with a birth, life span, and impending death?

There is a hypothesis that supposes that our planet (and others as well in the galaxy and Universe) is, in fact, alive. In a series of books published in the 1970s and 1980s, a British scientist named James Lovelock argued that Earth is a living organism, even renaming our planet Gaia, for the Greek goddess who drew the living world from chaos. He argued that the interplay of air, water, rock, and organisms that create oxygen and store carbon resembles the biological system of a body, and that Earth's peculiar balance would be recognized by any interstellar visitors as the product of life. Lovelock wrote:

Just as the shell is part of the snail, so the rocks, the air and the oceans are part of Gaia. Gaia, as we shall see, has continuity with the past back to the origins of life, and in the future as long as life persists. . . .

You may find it hard to swallow the notion that anything as large and apparently inanimate as the Earth is alive. Surely, you may say, the Earth is almost wholly rock, and nearly all incandescent with heat. The difficulty can be lessened if you let the image of a giant redwood tree enter your mind. The tree is undoubtedly alive, yet 99 percent of it is dead. The great tree is an ancient spire of dead wood, made of lignin and cellulose by the ancestors of the thin layer of living cells which constitute its bark. How like the Earth, and more so when we realize that many of the atoms of the rocks far down into the magma were once part of the ancestral life of which we all have come.

Lovelock's conviction that the Earth is actually a living organism has been embraced by some in the environmental and New Age movements, but it remains unaccepted by most scientists—including us—because the Earth as a whole doesn't reproduce, metabolize, and evolve (although parts of it do all three). Organisms need both energy and new physical material to grow, and while Earth constantly receives energy from sunlight, it (except for a small contribution of new meteorites) simply recycles the material that is already present.

However, the recognition that specific properties of the Earth that are required to maintain life are both self-regulating, and changing, is scientific fact. There are highly complex "life support systems" that sustain atmospheric compositions and pressure, planetary temperature, and even surface land features that are quite different from those of a lifeless planet. Accordingly, a new scientific discipline called Earth System Science has resulted from the original musings of James Lovelock. It is one of the main reasons that we have the temerity to suggest that the various ends of the world are indeed predictable.

As an example, consider Earth's remarkably stable temperature. We will explain in more detail why our Sun is getting hotter, but for now remember that its brightness has increased 30 percent since the creation of our planet. Venus, a hellish planet with a surface temperature sufficient to melt lead, gets 50 percent more solar radiation than our planet because it is closer to the Sun. Why hasn't Earth become a nightmarish Venus? What sets the planetary thermostat that allows water to remain between the freezing and boiling points, thus allowing life?

The answer is three cycles: plate tectonics, the carbon cycle, and the carbonate silicate cycle, or the weathering of stone. We must understand how these operate now to understand what will happen to our planet in the future.

PLATE TECTONICS IS A PROCESS THAT PRODUCES (AMONG MANY other things) the movements of landmasses that we call "continental drift." Nothing is more important in regulating Earth, because it helps sustain the balance of rocks, ocean, and air. In *Rare Earth,* we argued that plate tectonics is so important that it might be a necessary requirement on any planet for the evolution and maintenance of complex life. On Earth, plate tectonics—in which continents and ocean floor "plates" drift and grind against one another like a skin atop bubbling porridge—maintains surface temperatures to allow liquid water. As the early geologist James Hutton presciently suggested, we can imagine the slowly moving continents, spreading ocean floor, and the enormous, molten Earth convection cells far below that drive it all (none of which Hutton knew about in his day) as analogous to an enormous circulation system. But this circulatory system not only carries material from place to place, it changes it, helping to maintain the oceans and atmosphere so necessary to life.

To understand why Earth works so well today, and will eventually break down, we must understand plate tectonics. The continents are still drifting today, with the Atlantic Ocean imperceptibly widening and the Pacific shrinking. The squeeze of this shrinkage helps produce the volcanoes and earthquakes of the "Pacific Ring of Fire." The Himalayas were created when India crashed into Asia, and the Alps when Italy butted into Europe. Antarctica was once far enough north to have tropic vegetation, and Africa and South America once fit together like a jigsaw puzzle.

Our crust moves because the interior of the Earth is hot, and this heat comes from the slow decay of radioactive elements deep inside the Earth. As the heat rises toward the surface, it creates gigantic convection cells of hot, liquid rock in the mantle. Like boiling water, these massive cells of the viscous upper mantle rise, cool as they move parallel to the surface of the Earth, and then sink. As they move, they carry our planet's brittle crust with them. Sometimes this outermost layer of crust is composed of ocean bed. Sometimes it is the continents, the lighter granite and andesite "floating" atop the heavier basalt of the ocean floor. The rock that forms the foundation of the continents is lighter in color than the basalt because it contains more lightweight silicates, which tend to be white, due to the large amount of quartz in the rock. This continental rock is made in several steps, but a key ingredient is water, and its manufacture is one of several examples of how rock, water, and air interact with one another.

Plates are of varying thickness, their bottoms melting away when temperatures down there reach 1,400 degrees Centigrade. Ocean plates are about 60 kilometers thick, and continents average about 100 kilometers in thickness.

This bubbling stew is constantly making new crust. Hidden beneath the ocean waves are our planet's longest mountain ranges, incredible midocean ridges that circle the Earth like the seams of a baseball. These ranges are fissured down their middle and from these fissures, new basaltic crust vents in the form of magma. It boils and oozes and then rapidly turns solid in the near-freezing temperatures of the deep sea. As more magma wells up behind it, this new crust is pushed outward from these "spreading centers." Over many millions of years the oceanic crust moves away from its birthplace—the spreading centers—all the while being carried piggyback style on the convecting mantle beneath it. Like all journeys, however, this long ride must eventually end. Typically near its junction with continents, this spreading ocean floor sinks back down from the pull of gravity where the bubbling convection cell brings its slow return back into the depths of Earth.

It is at these subduction zones that continental mountain ranges are constructed. Some of these ranges are caused by buckling from the collision of ocean plate and continental lip, and some by the upward movement of hot magma that solidifies into granite or other magmatic rocks. In the United States, the ancient Appalachians are worn low, and the much newer Rocky Mountains are still high, because the Rockies were near the edge of the continental subduction zone in relatively recent times, were pushed up, and have not had as much time as the Appalachians to be eroded. This process has since moved westward, and the Cascade Mountains of Washington State are an example of recent mountain building. Glacier-clad volcanic peaks such as Mount Rainier, Mount Baker, and Mount St. Helens are direct evidence of the power and importance of subduction in creating mountain ranges. While other planets and moons in our solar system have volcanoes, they do not have mountain ranges like Earth's. This is clear evidence that only Earth has plate tectonics.

Earth's lighter continents float like gigantic rafts, and these rafts that are so important to the evolution of life are the product of water. As the deep ocean basalt created in the spreading centers moves away from its birthplace, it changes composition as water is added to the crystal structures of key minerals. These minerals become hydrated: the water molecules are added to the crystal lattices of minerals making up the basalt. This hydrated rock melts first as the basalt begins to descend again into the Earth, and the liquid that is produced rises back to the surface of the planet, typically cooling as granite or andesite, which floats above the ocean floor. This becomes the backbone of the continents. Because continents are relatively light, they can never be sunk. Nor can they be

THE LIFE SPAN OF HABITABLE PLANETS · 57

destroyed. Continents can be split, fragmented, and shuffled about, but their basic volume can never be decreased. In fact, they have been growing since the formation of the Earth.

This would seem to be counterintuitive, since the ocean crust is continually enlarging due to seafloor spreading. Yet as we have seen, ocean floor can sink and be remelted back into the magma, while continents cannot. Meanwhile, the newly hydrated minerals are constantly floating upward in the form of lava from volcanoes and new granitic and andesitic magma. It has been estimated that the volume of continents increases by 650 to 1,300 cubic kilometers per year! The old adage to "buy land because they're not making any more of the stuff," is not, in fact, strictly true. Geologists have suggested continental crust formed at even faster rates in the past, when the Earth was probably even more volcanically active because of higher heat flow from its interior.

As new crust erupts from volcanic fissures and old crust is subducted into the mantle at the lip of continents, it causes chemical change through new mineral formation, heating, and the liberation of gases and water vapor. These help maintain a constant planetary temperature by removing some elements (through subduction) and reintroducing others (through volcanic eruption). We might analogize plate tectonics with the physiological system that allows mammals and birds to maintain a constant body temperature that is neither too hot nor too cold.

Yet this is but one of the contributions of plate tectonics to the world as we know it. It also acts as a global recycling system of elements essential for life, such as phosphates, nitrates, and carbon. The latter is not only an essential building block of all organisms, but its cycle regulates how hot or cold the Earth's surface gets.

CARBON COMBINES WITH OXYGEN TO FORM CARBON DIOXIDE, a "greenhouse gas" that warms a planet's surface by retaining the Sun's radiant heat. Methane and water vapor are also highly effective greenhouse gases. Human combustion of fossil fuels such as gas and coal releases stored carbon into the atmosphere in the form of carbon dioxide, or $CO_2$, and it is the buildup of this gas that has led the majority of scientists to conclude that human-caused global warming is under way. Over the Earth's billions of years of history, the release and storage of carbon has helped keep its temperature—most of the time—at habitable levels. The carbon cycle is thus critical to the life span of habitable planets.

The movement of carbon from the inorganic world to the live, or organic, world has been described by Lee Kump, James Kasting, and Robert Crane in their 1999 textbook, *The Earth System*. They begin with one $CO_2$ molecule, floating in the atmosphere. Over a decade or so it is wafted around the surface of the Earth by the moving, turbulent atmosphere. When it finally encounters a plant, it passes through one of the small stomata openings found in the leaves and has its two oxygen atoms stripped away, to be replaced by hydrogen, nitrogen, and other carbon atoms through chemical bonding. It becomes one of the galaxy of atoms making up the material framework of the plant, and is thus changed from inorganic to organic, or from an atmospheric molecule to part of a leaf.

When autumn comes the leaf falls, disintegrates, and becomes incorporated into the soil. Our carbon atom is eaten by a soil bacterium and transformed by a chemical reaction back into carbon dioxide that the bacterium exhales. Once more, it escapes into the atmosphere. The same liberation could have happened if a deer ate the leaf. On the atomic level the air becomes plant and the plant becomes air, over and over and over again. As solid as they are, plants are basically a manufactured combination of water, air, and sunlight.

Variations of this cycle might be repeated an average of five hundred times until a different fate occurs. In this case the soil containing our carbon atom is eroded and transported by water to the sea. There, it might again be consumed by an organism, but in this case let's assume it escapes that fate and is instead buried by sedimentation. As more and more sediment falls the carbon atom is interred so deeply that it is now in a world of virtually no oxygen, and thus exists in an environment that can no longer be perturbed. Nothing eats it. Nothing exhales it. It is entombed by rock, no longer contributing to the greenhouse effect in our atmosphere. This molecule of carbon has been taken out of play.

Millions of years later, however, plate tectonics thrust the sedimentary rock into high mountains. Now wind and water go to work, and the rock is eroded. The carbon atom is liberated, bonds with atmospheric oxygen, and once again forms the greenhouse gas carbon dioxide!

The various fates of our carbon atom profiled above suggest that there are numerous "holding tanks," or reservoirs, where carbon is stored. Balancing those reservoirs is vital to regulating our planet's temperature. The atmosphere holds 800 billion tons of carbon, the

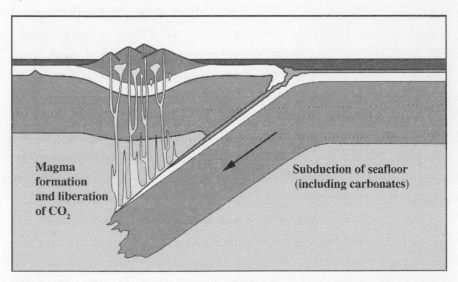

Magma formation and liberation of $CO_2$

Subduction of seafloor (including carbonates)

Plate tectonics and its associated subduction of seafloor material beneath continents is a critical factor in maintaining Earth's long-term habitability. The process not only produces mountain ranges but also recycles carbon dioxide back into the atmosphere, driving our planetary thermostat and providing sufficient carbon dioxide for the survival of plants.

soil twice that, and the oceans fifty times as much. But by far the greatest volume is stored in sedimentary rock and limestone. Plate tectonics is a conveyer belt carrying this carbon deep into the mantle, and then releasing it through volcanoes in a cycle that has helped keep Earth's temperature balanced. While the planet in the past has been hot enough to make high latitudes tropical and cold enough to cause Ice Ages, it has never become as hellish and uninhabitable as Venus or as frozen and hostile as Mars.

The detailed mechanism of how carbon is balanced is called the "silicate-carbonate geochemical cycle." It balances inorganic reactions taking place deep in the Earth with interactions between the atmosphere and the surface of the Earth, and it requires the help of plants and animals. Which brings us to our third key cycle that helps keep atmospheric carbon dioxide levels essentially constant— and hence keeps the Earth's surface temperature relatively constant—for the long scale of geologic time.

TWO QUITE DIFFERENT PROCESSES ARE KEY. THE FIRST IS CARbonate precipitation. If calcium—used widely by animals to build shell or bone—is combined with carbonic acid under the right temperature and pressure conditions, it can combine to form calcium carbonate, or limestone. This is one of the most common of all sedimentary rocks, and its creation removes carbon from the atmosphere. In fact, the more carbon dioxide in the air, the faster limestone formation will occur if sufficient calcium is available.

Where does the essential calcium to make limestone come from? Ultimately, from igneous, sedimentary, and metamorphic rock brought to the surface by volcanoes through plate tectonics. The same eruptions that are spewing carbon dioxide into the atmosphere are delivering the chemical necessary to ultimately remove that same carbon, and thus keep the atmosphere in balance. As Louis Armstrong sang, what a wonderful world.

But where to get carbonic acid? This comes from the second necessary process, the weathering of a class of rocks known as silicates, such as feldspar and mica. As we have seen, common rocks such as granite are largely composed of silicates. When silicate rocks erode, the by-products can combine with other compounds to produce calcium, silicon, water, and the all-important carbonic acid, which is needed to create the limestone.

We've already noted that the more carbon dioxide there is in the atmosphere, the faster limestone can form to remove that same carbon dioxide. Similarly, the higher the concentration of the greenhouse gas called carbon dioxide, the warmer the planet gets. A warmer world generates more evaporation, and thus more rain and wind. The more rain and wind, the faster weathering occurs. The faster such erosion occurs, the faster carbonic acid is liberated—and thus, again, the faster limestone can occur and the faster carbon dioxide is removed from the atmosphere, ultimately cooling the planet back down. As the planet cools, weathering slows, carbon rises, and the Earth heats up again. In a sense, our planet is self-regulating. And this sustains a habitable temperature.

Here we have a wonderful partnership. Animals such as coral are harnessing calcium. The roots of plants exude an acid that helps to break down rocks, accelerating weathering by the wind and rain generated by the atmosphere of oceans, creating the acid necessary to convert the calcium to limestone. All combined are working together to take excess carbon dioxide out of the atmosphere and bury it in "reservoirs" of rock within the Earth, and thus balance temperature. Plate tectonics simultaneously helps remove carbon, through subduction, and reinject it through volcanoes. If this cycle didn't operate—this "circulation of rocks"—there would be nothing but bacteria on this planet because Earth would have overheated or gotten too cold. In fact, it is such an exquisitely balanced and self-regulating system that Lovelock's argument that the planet can be "alive" is understandable.

The actual details of this recycling system can get quite complicated, hinging on how much land area is available to weather, how much plants are contributing to weathering (because their roots help break rock apart), how fast the seafloor is spreading and subducting to generate volcanoes, and so on. The fine balance can also be temporarily upset. During global warming, ocean water actually releases more carbon, accelerating the warming into what can become a "runaway" greenhouse effect. During an Ice Age, plankton actually take up more carbon, accelerating the cooling. Because of this, our planet has had many periods where it was too hot or too cold to be ideal.

Still, our overall point is simple: rocks, air, water, and life have in general combined up to now to keep our planet's temperature from going to extremes severe enough to kill all life, as happened on planets such as Venus and Mars. Plate tectonics, the organic to inorganic carbon cycle, limestone formation, and weathering all work together to make Earth habitable. Will it always be so?

A TEAM OF SCIENTISTS (BOB BERNER, TONY LASAGA, AND BOB Garrels) has tried to answer this question by using a mathematical model to look 600 million years into the past, or roughly the time that animals and higher plants have existed on planet Earth. Their resulting graph shows several interesting trends, the most important being an overall, long-term decrease in carbon dioxide. In the period just before the Cambrian Explosion of new animals, $CO_2$ levels were about fifteen times higher than present-day levels. Over the subsequent 100 million to 150 million years, carbon dioxide gradually increased through a series of fluctuations to more than *twenty* times present-day values. Then, about 400 million years ago, the most remarkable thing happened: $CO_2$ levels dropped— markedly. The reason for this drop seems clear. The time interval of about 400 million years ago coincides with the rise of vascular land plants.

As land plants began to cover the planet and evolve from the first sparse twiggy forms to forests of trees, enormous changes affected the planet. Great quantities of carbon began to be locked up as rotting vegetation, and eventually coal and oil. Soils became deeper and richer. And as the green spread, the fine balance between the amount of carbon held in the atmosphere and that held in the soils, oceans, and rocks of the planet began to change. Carbon dioxide levels began to drop. Plants sucked it from the atmosphere to grow. Their soil became a carbon reservoir. Plants began to increase weathering rates of silicate rocks as their roots secreted acids to obtain nutrients and a foothold, thus allowing ever more limestone to form. Huge amounts of carbon dioxide were locked away in coal, oil, and gas beds.

This decline in $CO_2$ levels appears to have continued over the past 100 million years. Much of this decline may have been driven by tectonic events, most notably the geological uplift and subsequent weathering of the Himalayan Mountains. Because this largest of Earth's mountain ranges is composed largely of silicate rocks, and because its extraordinary uplift (due to the chance collision of the Indian tectonic plate with Asia) created the thickest continental crust on the planet, this single event seems to have markedly changed atmospheric $CO_2$ composition, and, with it, the Earth's climate. As the Himalayas weathered, they allowed the various chemical carbonate cycles to remove atmospheric $CO_2$ faster (through limestone formation) than volcanic system could replace it. Over the past 60 million years this event, coupled with the further spread of plant life across ever-growing land areas, drove the levels of $CO_2$ to historically low levels.

What has this meant for climate? The history of our planet's temperature is not easily studied. There are no direct "paleo-thermometers" that give some mean global temperature at any given time. While there are a few ways of measuring ancient temperatures from extinct organisms—such as studying the ratio of isotopes of oxygen as recorded in sedimentary rocks—these records are for individual locales rather than for the planet as a whole, and applicable mainly over the past 100 million years. That represents just 2 percent of the Earth's history. For the other 98 percent of time we must rely on indirect evidence from the geological and paleontological records.

Among such clues is the presence of specific sedimentary rocks that are indicative of ancient climate. For example, sedimentary rocks known as evaporate (such as ancient salt deposits) are indicative of heat. Glacial deposits tell us of ancient cold. Fossils are also of great importance, for specific types of organisms are often useful in interpreting ancient climate. Fossil soil types are similarly useful, for soils are highly climate sensitive.

Using such methods, paleoclimatologists have arrived at an accepted temperature record of the past 540 million years, the time of abundantly skeletonized fossils. This interval of time, which is composed of the Paleozoic, Mesozoic, and Cenozoic eras, has shown periods both warmer and colder than the present-day mean temperature, which is about 15 degrees Centigrade, or 59 degrees Fahrenheit. But the temperature variation has not been large—only as much as 10 degrees C either hotter or colder that today's mean temperature any time during this long roll of history. We thus may have had an Earth with a mean temperature as hot as 25 degrees C, (77 degrees F) and as cold as 5 degrees C (41 degrees F). Neither of these extremes would have imperiled the continued existence of animals and plants on the planet.

What of deeper in time, back before the common presence of animals? Here the record is much more obscure, and the degree of consensus about the record of mean global temperature much more tenuous. The recent recognition that the entire planet froze over about 2.3 billion years ago, and again about 700 million to 600 million years ago, suggests that there have been greater swings of global temperature in the period before plants and animals arose. Studies from the 1980s suggest that the planet cooled dramatically from the 80 degrees Centigrade, or 176 degrees Fahrenheit, it broiled under some 3.8 billion years ago when planetary bombardment ceased. By 3 billion years ago, the mean temperature had fallen to 40 degrees C (104 degrees F), and by 2 billion years ago to 20 degrees C—with global temperature never exceeding 30 degrees C subsequent to that time. This interpretation suggests that temperature had little to do with the evolution of life subsequent to about 3 billion years ago. However, this view is no longer universally accepted. Oxygen isotope records derived from a series of pristine chert—rocks that had been deposited well before the advent of skeletonized animals—give a very different story, suggesting global temperatures stayed warmer for much longer periods of time: 70 degrees C at 3 billion years ago, 60 degrees C at 2 billion years ago, and about 40 degrees C as late as 1.5 billion years ago. If this is the case, then evolution may have taken so long to get going because the planet (including the oceans) was simply too warm.

Whatever the pattern, one thing is striking. Once life began to develop in complexity and emerged from the sea, temperature appears to have been stabilized. Yes, there were Ice Ages, and yes, there were periods when most of the planet was tropical. Still, life conspired—by weathering rock with its roots, building shells of its limestone, and storing and exhaling carbon—to keep things in a habitable range.

Prior to the evolution of vascular plants in the Devonian period, some 400 million years ago, there must have been a very different rate of chemical weathering on land, and hence far higher atmospheric carbon dioxide and planetary temperatures. Although some researchers have suggested that there was sufficient fungi and algae on the Earth surface prior to this time to affect chemical weathering rates, most agree that it was the evolution of root systems that was the most significant change. Experiments show that the presence of vascular plants accelerates weathering from four to ten times. All of us are familiar with the cool respite of a tree's shade. It is interesting to contemplate that over long eons of time, even more important has been the cooling effect of the plant's vigorous roots. Add to this the carbon that plants store in coal, soil, woody debris, and so on, and it is perhaps no surprise that atmospheric carbon dioxide has dropped twenty times. This reduction made possible the Ice Ages. And, as we shall see, a continued drop will have even more dire biological consequences.

Which brings us back to Ruth Ward.

The death of Ruth was initiated by the breakdown within one particular system, her digestive system. Ultimately that incident impacted upon her many other systems, including her circulatory, respiratory, nervous, lymphatic, excretory, immune, and hormonal systems. Each of these linked systems is necessary to keep a human functioning, and the breakdown of one always impinges on the others. Sometimes these breakdowns are fixed. Other times they spell death.

The same is true of planets. The circulations of carbon, nitrogen, sulfur, phosphorus, and various trace elements are necessary to sustain life, and these in turn are affected by geologic, atmospheric, and oceanic systems. By looking at records of the past such as fossils, rocks, ice cores, sea-bottom cores, and so on, we can understand how our planet has operated up to now. By knowing how these systems work, we can understand what may happen when one or more of them cease to work. If we can predict how the various planetary support systems will change in the future, we can predict the ends of the world. Our time machine is nothing more than a series of predictions of slight and seemingly mundane changes in atmospheric gas levels, the flow of nutrients and the flow of energy from the Sun, and the heat generated within the Earth by radioactive decay.

There's a difference between a human's life and the life of our planet, however. Ruth Ward, born in 1916, aged gracefully but never resembled her youth. Hers was a one-way trip. Planets have a different trajectory: the Earth appears to be on a round trip of sorts. If you take a cannon and shoot the projectile straight up, it climbs to a certain height, slows, stops, and then falls back to the ground. Our planet's trajectory is similar. It started as a very hot, oxygen-free world. Water, air, plants, solar energy, and plate tectonics created the conditions for evolution to its present state, and so most readers might assume that the cannonball of biological complexity is still arcing upward. The authors believe, however, that the cannonball has already began to drop, and that Earth has already started a return to a hot world where life becomes less diverse, less complicated, less abundant through time. The last life may look much like the first life—a single-celled bacterium, survivor and descendant of all that came before. As a result, our Earth is likely to relive many aspects of its past: including its very recent past.

· **4** ·

# THE RETURN OF
# THE GLACIERS

OUR FIRST STOP IN THE FUTURE IS NOT MILLIONS OF YEARS from now, but only thousands. While in terms of human life it seems inconceivably far away, from a planetary perspective it is but a blink forward in time. This future is so close to us that the planetary habitability systems profiled above will be little changed: the Sun will be no brighter, the amount of carbon will be quite similar, even the rate and nature of the tectonic recycling systems will be no different from out present day. Our world is not threatened with imminent death. But it is about to catch a huge cold, one that will prove a severe trial for us humans, and for the rest of the biosphere as well. We start here to show how even slight perturbations in the global thermostat can— and will—have huge effects on climate. The other result of glaciation is a reduction of the amount of life on the planet. But can we be sure that there were so recently thick walls of ice over huge land areas?

THE SIGNS NEAR OUR HOME ARE EVERYWHERE, ONCE YOU KNOW what to look for. The huge piles of gravel lining Puget Sound. The linear hills, lakes, and fjords that run north and south. The huge boulders left, like discarded litter, hundreds of miles from their mountain of origin. The scratches on harder rocks like gigantic sandpaper lines left by malevolent giants. It tells us that not so long ago, great rivers of ice flowed across the land.

The place where we live and work, the Pacific Northwest, is a most frustrating habitat for a geologist, for there is virtually no bedrock in the entire Puget Lowland. Once this land was like any other, with outcrops of ancient sedimentary, volcanic, and meta-morphic rock that proclaimed a long geologic history. But then the glaciers came, first covering the landscape with ice and then leaving a residue of gravel. The glaciers scoured or buried the older land-scape as if by whim, often slicing huge rents in the original bedrock, and thus either bringing great foreign boulders from the north or carrying local rocks farther south. As recently as fifteen thousand years ago, a mile of ice sat upon our future home sites in Seattle. It is by understanding this past that we can imagine a truly ominous future. Despite well-justified concerns about present-day global warming, we believe a far greater catastrophe looms at the other end of the temperature spectrum. Come with us to the Seattle of fifteen thousand years from now.

IT IS BITTERLY COLD, WINDY, GRAY, AND DRY: THE MOIST MARINE air we remember has been sucked away into ice. A glacial wall has replaced the downtown skyline of our city. Seattle's rubble ema-nates from the glacier's end, along with muddy water. To the west, the ground dips lower and more steeply than we remember into new empty canyons. As in the past Ice Age, sea level has dropped 400 feet and Puget Sound is largely emptied of water.

A large band of jocular kids is stepping up onto the glacier's edge and boldly walking across the dirty ice, leaving their alarmed teach-ers in the lee. The noisy pack runs and skids, pointing and shout-ing, muffled in their warm clothes against the searching cold and keening wind. They clatter across ice that now stood on the ancient watercourse of the Duwamish River, and the more knowledgeable name the finger of concrete and steel with a flying saucer on top that just protrudes from the glacial ice in the distance. A cry goes up: "The Space Needle! There was once a great city all around it!" One of the listening children frowns. How gullible did they think he was? A city beneath the land of ice that he knew stretched from this point to the North Pole?

The shabby adults who follow behind, keeping a wary eye on the ice and their offspring, envy their children's joy and innocence. The parents know the grim reality this ice has brought: meager crops, no more gasoline, and the specter of starvation. Books said that there were once apple orchards here, a long time ago, amid an ancient era worried about something called global warming. The fools! If only they could have kept the world "globally warmed" a little longer. But once all of the oil and coal were gone there was nothing left to keep the glaciers at bay, and from the North the walls of ice had once again flowed over the land, squeezing human civilization into the warmer climes of the mid- to low latitudes. Again. This band would like to retreat there, but the peoples of the tropics are already crowded into their own countries and dealing with their own desolation. The once great rain forests of Amazonia, Asia, and equatorial Africa are changing into savannas, and the once endless steppes and grassland plains of the midlatitude world are dust-bowl deserts. There is no room or food for the peoples of the North. Unlike the last Ice Age of the cavemen, there is no longer any place on Earth not already teeming with humans.

AT THE START OF THE TWENTY-FIRST CENTURY, THERE IS A PLANetary consensus emerging that rapid global warming of the planet, caused by human air pollution, poses a significant global threat. A warming world will bring about rising sea level, changing plant communities, and the migration of tropical diseases into temperate regions. There may be an increase in the violence of storms, more droughts and floods, and disruption of agriculture. As Earth heats up, wars over food, water, and habitable land may increase.

As dire and as real as this threat is, it is very near term by the time scales we are dealing with in this book. Yes, the atmospheric concentration of carbon dioxide from the burning of fossil fuel such as coal, gas, and oil has increased 30 percent since the start of the Industrial Revolution, to the highest level in 450,000 years. The greenhouse gas methane is up 145 percent. Nitrous oxide is up 15 percent. Evidence that this is causing real and dramatic change in the world is rapidly mounting. The decade of the 1990s was the warmest on historical (not geologic) record. The Arctic icepack has thinned 40 percent in the past twenty years. The Alps have lost half their glacial mass since the nineteenth century, and the snow of Africa's Mount Kilimanjaro may be entirely gone in fifteen years. Shrubs are growing on Arctic tundra. Mosquitoes are found at higher elevations in the tropics, carrying diseases that are pushing bird species to extinction in places like Hawaii. And this may be just the beginning. The International Panel on Climate Control, representing twenty-five hundred scientists, says average global temperatures could climb from 1 degree to 6 degrees C in the twenty-first century. A rise of 6 degrees could shift climate zones 500 miles toward either pole, disrupting agriculture, watersheds, and snowpack.

As catastrophic as this could prove to be, it is but a momentary event in the long history of our planet. In a few centuries, most fossil fuel will likely be exhausted. New energy technologies will have to be developed. The pumping of carbon dioxide into the atmosphere will slow, then cease, and natural mechanisms—outlined in the previous chapter—will begin to lower greenhouse gas concentrations. From the perspective of our lives and human history, centuries of global warming is a long time: as long as countries or empires often exist. But from the perspective of planetary history, this warming will be a brief interlude before an inevitable return to the more persistent age we inhabit. That age is an age of ice.

That we are in a 2-million-year period of advancing and retreating ice is odd. The previous major ice age was 260 million years ago, when the Sun was less energetic than it is today. But even though we are increasing carbon dioxide today, the rising heat of the Sun has been outstripped over millions of years by a long-term decrease in atmospheric carbon dioxide. As discussed in the previous chapter, carbon dioxide levels have declined to 5 percent of what they were before the rise of plants and animals. Much of this decline is due to the sequestration of carbon by plants both on land and in the sea, and more may be due to the ever-enlarging continents that are being built by plate tectonics. As this big landmass weathers, it draws carbon dioxide out of the atmosphere. The result has been to tip Earth toward an Ice Age.

There is abundant scientific evidence that we live today in world that is quite atypical of planet Earth over most of its past history—and probably atypical as well of much of Earth's future history. It has been cold, unusually cold, for 2.5 million years, and at the same time the gentle green cradle we regard as normal is, in fact, very temporary. Human civilization has arisen in a brief "interglacial" that has lasted only about twelve thousand years and may already be ending. Human-caused global warming may stall the return of the glaciers by a few centuries, but it will not prevent them. The wild swing from overly hot to overly cold will only make the whiplash change in climate seem more severe. While the glacial return will not threaten the existence of life on our planet, it will result in a many-thousandfold reduction in the biomass, or volume, of living plants and animals. This will effectively end the world as we know it—and potentially end human civilization as well. Humans will have to abandon their planetwide occupation and be content with a more precarious existence in the mid- to low-latitude regions of the Earth.

The Pleistocene epoch, or the Ice Age as it is more popularly known, began when more snow fell each winter than melted in the spring. Year by year this excess of snow and ice caused the formations of glaciers, which slowly crawled southward. Eventually continental glaciers began to coalesce and merge with mountain glaciers, uniting in unholy matrimony to grip the land in glacial ice and glacial winter.

By no means was the entire planet gripped in ice. There were still tropics, and coral reefs, and warm sunny climes pleasant the year-round. But probably no place on Earth save the deepest sea bottoms were unaffected because when the global climate changed, so did wind and rain patterns. Even those places far from the ice were climatically changed, perhaps colder, occasionally warmer, and often quite drier. Gigantic, cold deserts and semideserts expanded in front of the advancing ice sheets, while regions normally dry, such as the Sahara Desert of northern Africa, experienced increased rainfall. Conversely, the great rain forests covering the Amazon Basin and equatorial Africa, regions of relative climatic stability for tens of millions of years prior to the onset of the Ice Age, experienced a pronounced cooling and drying. Large tracts of jungle retreated into pockets surrounded by wide savannas.

In North America the southward advance of the ice halted in midcontinent, the maximum extent of the latest glaciation occurring eighteen thousand years ago. To the north, most of the land was uninhabitable ice. To the south, drier regions of shifting sand produced huge deserts and sand dunes. It was surely an extraordinary time in the history of our planet, but not unique. Glaciation has affected significant portions of the our planet many times in the remote past, such as during the Precambrian era a billion years ago and during the Permian period of 260 million years ago. Yet the last episode, which ended only twelve thousand years ago, was one of the most intense.

The glaciers changed the nature of life on Earth, and in many regions the geography of the planet itself. In Europe, the expanding and retreating ice sheets carved the fjords of Scandinavia as well as many of the geomorphic features of northern Europe. In North America, the ice gouged Puget Sound in Washington State and the great Inside Passage stretching from southern British Columbia to Alaska. In the midcontinent, the Great Lakes were created, while in Asia huge lakes such as Lake Baikal were similarly created by the action of the moving ice. Huge lakes dammed by ice were built and released, creating monstrous spillways when they eventually broke. Finally, the retreat of glacial sheets produced huge piles of gravel and debris spread over large expanses of all the northern continents. The ice sheets were more than a mile thick in most places.

As dramatic and planet sculpting as this last Ice Age was, it is by no means the first, and by no means the worst. Some 2.5 billion years ago, and then between 700 million to 800 million years ago, before complex animals existed, there were Ice Ages so extreme that the Earth was covered from pole to pole with ice. Even the oceans were frozen in these events, which have come to be known as "Snowball Earth." These anomalous events occurred when the Earth's system of temperature control seems to have gone haywire and the oceans froze all the way down to the tropical latitudes. The freeze-overs lasted only a few million years and they were merely a blink of the eye in Earth's billions of years of history, but they were dramatic and their effects may have drastically altered the evolution of life. They could have stimulated the explosion of animal life at the start of the Cambrian, and they may have played a role in earlier times as well. Alien astronomers viewing the planet from afar would have seen our "pale blue dot" (in the phrase of the late astronomer Carl Sagan) turn white when mean global temperatures dropped to −50 degrees C, and then turn pale blue again when the ice melted and global temperatures approached +50 degrees C, a wild swing indeed. Another severe Ice Age occurred 400 million years ago, and another between 300 million and 270 million years ago.

But from that Permian period to the near present, a period of

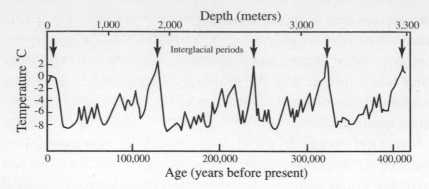

A 400,000-year temperature record derived from measurements on a deep ice core drilled at the Russian Vostok station in Antarctica. Four 100,000-year glacial cycles are seen with the temperature peaks marking the comparatively brief and warm interglacial periods. Note the unusual width of our current interglacial period, which has enabled the rise of civilization.

more than a quarter billion years, our planet was spared these harrowing episodes. Dinosaurs evolved and ruled, and then mammals succeeded them in domination. Then, just 2.5 million years ago, the climate finally changed and the times of ice returned. We are once again in a glacial age, and don't know when it will end.

IT HAD LONG BEEN POSTULATED THAT THE ICE AGE WAS COMposed of four separate glacial advances and retreats during these 2.5 million years. But improved dating techniques that measure the decay of radioactive isotopes now indicate there have been at least eighteen separate ice advances and retreats. Overall Earth has remained cold but, like clockwork, brief interglacial warm periods have occurred roughly at hundred-thousand-year intervals. Ominously, the severity and size of the glaciers produced during each cycle has been increasing through time.

As might be expected, the last glacial advance and retreat left the newest and least-disturbed geologic record. Known as the Wisconsin glaciation in North America and the Wurm in Europe, it ended about twelve thousand years ago. Ice covered Canada, the northern United States, and northern Europe, and so much seawater was imprisoned in the growing glaciers that sea level was four hundred feet lower than it is now. This alone created extraordinary circumstances.

If sea level were to drop that much again, Long Island would no longer be an island and the barrier islands of the Gulf Coast and the Carolinas would become part of the mainland. The coast would expand by tens of miles in most places. Yet even these changes would be minor compared to other parts of the world. Alaska would once again be connected to Asia. England and Ireland would be part of a vast peninsula extending out from France. Vast regions of the Mediterranean would go dry. India and Sri Lanka would link, Japan would be connected to China, and Australia and New Guinea would become one large continental block, separated from the Asian mainland by narrow straits. Just eighteen thousand years ago, this is what the world looked like—and may look like again, even a few thousand years from now.

At the climax of a new ice age, glaciers would extend to New York City and central Europe and the Atlantic Ocean would be clogged with icebergs. The ice caps would be up to three kilometers thick, or eight times the height of the Empire State Building. Wind speeds off the glacial cap would reach three hundred kilometers per hour, creating a wasteland of dust and sand along the glacial fronts. A treeless, tundralike landscape would extend for hundreds of miles south of the glaciers themselves. Still farther south, great deserts would form.

Why does the Earth undergo these radical and catastrophic cooling events? To put the Ice Ages in context we need to take a longer view back through time. Abundant evidence paints a picture of a much warmer world some 60 million to 70 million years ago. Sea level was much higher than in the present day, and global temperatures were remarkably constant from pole to equator. It was a lush, tropic jungle world during a time when the continents were more closely massed together and when several of the major mountain chains, such as the Himalayas, Cascades, Rockies, and Andes, were either not yet developed or still building and nowhere near their present height.

Since that time, however, there has been a long, slow drop in temperature caused by a combination of lower atmospheric carbon dioxide, an increase of landmass in the northern hemisphere, and changes in ocean circulation patterns. Carbon dioxide levels today may be only a tenth what they were 60 million years ago, shortly after the time of the dinosaurs came to an end. By 2.5 million years ago, conditions had changed sufficiently for ice ages to begin again.

Like so many popular usages, the term "Ice Age" is quite imprecise. Sometimes it is used to refer to an entire cold period, such as the past 2.5 million years, and sometimes to the periodic glacial advances, like the Wisconsin-Wurm, within that period. A better term for the former is a glacial age, and for the latter a glaciation. We are currently in a glacial age, and we are facing another glaciation.

The past ten thousand years have been a time of extreme warmth compared to the past hundred thousand years. The last such warmth was approximately 120,000 years ago, at the end of another glaciation. At that time, the Earth was about as warm as our own time for fifteen thousand years and then plunged back into ice. But a more detailed examination of this preceding interglacial shows a troubling difference: even during that time of warmth, the Earth's climate underwent substantial temperature changes in as little as ten years or less. In fact, a core of ice drilled near Russia's Vostok base in Antarctica shows that all previous interglacials prior to our own were times of major swings in climate. Why is our own interlude so stable? More important, how long will this stability last?

To answer this we must understand what causes glacial ages and glaciations, an issue of continued research and debate. We know these events are caused by cooling, but the difficult part is figuring out the precise reasons that the cooling occurs. The history of climate change is recorded in a variety of geological clues. Some of these data come from cave formations, where the slow accumulation of calcareous deposits on cave floors can be teased apart and analyzed using sophisticated laboratory machines. This yields astonishingly precise measurements of ancient temperatures. Accumulated deposits on the seafloor and on the bottoms of lakes also yield a splendid record of past climate. But one source of information surpasses all others: the cores of ice extracted from old sheets of ice in Greenland and Antarctica. These ice-core records not only give very high resolution temperature data for the past four hundred thousand years, but they also yield information about the frequency and strength of ancient storms and monsoons, the amount of dust in the atmosphere, the extent of the world's wetlands at any given time, and even the amount of salt in the ocean. They show quite clearly that over the past hundred thousand years there was a ninety-thousand-year slide from warm conditions—not unlike those of today—into a stark, bleak glacial interval, followed by a rapid warming that brings us to the present day. But these cores also show that the past ten thousand years have been atypically calm and constant in terms of global temperature and climate. Our "norm" is not normal, according to all the available evidence. It is the "why" of this fluctuating climate that has most intrigued climatologists.

Much of the answer seems to be the trace amounts of gases in our planet's atmosphere known as the greenhouse gases: carbon dioxide, methane, water vapor, and so on. When carbon dioxide levels dropped over the past 70 million years, this decreased the amount of solar heat trapped by the greenhouse effect and cooled the average temperature of our planet. Another answer is continental drift. When continents move toward the poles, they provide a platform on which the thick ice sheets form. The ice of the Antarctic ice cap, for example, averages nearly two miles thick because it sits on a continent. The ice of the Arctic Ocean, where there is no land, is only a few meters thick. Still another factor may be changes in ocean circulation. We know that when North and South America were joined at Panama several million years ago, a wave of extinctions occurred as animals from the two continents mingled. It also shut off easy circulation between the Pacific and Atlantic Oceans, and some theorize this changed weather patterns.

Finally, a critical reason may be Earth's orbit around the Sun. This latter idea was first proposed by Yugoslav astronomer Milutin Milankovich, for whom the orbital theory is named. He proposed that three variations in the Earth's orbit change the amount of solar energy that hits our planet, and thus the overall global temperature.

The first variation is that the Earth's orbit around the Sun changes from circular to a pronounced ellipse over a period of about ninety-five thousand years. Because of this ellipse today, Earth is closest to the Sun on January 3 and farthest away on July 4. And since most of the world's land is in the Northern Hemisphere, and because the Sun is more distant in the northern summer, we live in a period in which summer snows may last longer than the long-term norm. During a glaciation, they accumulate faster than they melt. The shape of the Earth's orbit at present could be one reason why this happens.

The second variation is that while the Moon keeps the tilt of the Earth's axis relatively stable compared to planets such as Mars, it still varies between 21.8 degrees and 24.4 degrees over a period of forty-one thousand years. This tilt is what causes our planet to have summer and winter, and the greater the tilt, the greater the difference between summer and winter—and the greater the chance that snow and ice could begin to accumulate.

The third variation is that the Earth also wobbles around its axis over a period of twenty-two thousand years: an oscillation like that of a spinning top that scientists call precession. This changes the season when the Earth is closest to the Sun. Today, the Northern

Hemisphere has summer and the Southern Hemisphere has winter when our planet is farthest from the Sun. About eleven thousand years from now, the situation will be reversed.

Each of these factors influences how much sunlight strikes the Earth at a particular time. Milankovich's genius was in figuring out how to combine all three at once to make predictions. Looking backward six hundred thousand years, he argued he could correlate glaciations with these cycles of orbital change. While ignored in his own time, he was vindicated in the 1960s when detailed climate records from deep-sea cores showed that there had not been four glaciations but at least eighteen, in line with his predictions.

Still, riddles remain. The changes in sunlight are slight; how could they make such an enormous climate difference? And why do ice sheets melt much faster than they grow? A glacial advance that has grown for tens of thousands of years can retreat in as little as a thousand. Why?

More recently a new possibility has arisen: every hundred thousand years or so, Earth passes through a ring of cosmic debris and dust that shades the planet and blocks enough sunlight to perhaps trigger a glaciation. This idea, proposed by Berkeley physicist Rich Muller, is still being debated.

The intricacies of the many linked systems that produce climate and climate change make accurate predictions about the future difficult. While we can forecast when Milankovich-induced changes in solar energy will occur, current scientific models cannot yet explain exactly how these long-term changes tip climate from one state to another. A different method of predicting the future is simply to look at the length of past interglacials.

Ice-core records and deep-sea paleontological and isotopic records indicate that over the past eight hundred thousand years, the interglacial periods—the warmer times between the much cooler glacial intervals—have lasted on average for half a twenty-two-thousand-year precessional cycle, or about eleven thousand years. The current interglacial has already lasted more than eleven thousand years, and some records suggest that we have been in the warm period for as much as fourteen thousand years. Does this mean that the glaciers are advancing at this moment? The answer to that question is a decided no, because precession is not the only orbital variable that affects climate. Records show that between 450,000 to 350,000 years ago there was an interglacial stage that lasted much longer than eleven thousand years. This interglacial coincided with a time when orbital eccentricity was at a minimum. Just such a pattern of minimal orbital eccentricity is under way at this time, suggesting that the present interglacial could continue for thousands or perhaps a few tens of thousands of years into the future. Or it could end at any time.

How long will the present icy period as a whole, already under way for 2.5 million years, persist? One prediction was made by R. Chris Wilson, Stephen Drury, and Jenny L. Chapman in their book *The Great Ice Age,* published in 2000. They predict that the present interglacial should end within a few thousand years at most, to be followed by a drop of global temperature by as much as 10 degrees C for the next eighty thousand years. Such a cooling would be sufficient to cause a severe new glaciation. Would this new onslaught of ice come gradually or suddenly?

Again, we turn to the past and learn that the preceding interglacial ended very abruptly, and with little warning. In the parlance of the climatologists who study these records, "someone" threw a switch, and the climate tipped over. The first evidence comes from Atlantic Ocean sediments, which show large amounts of rocks and cobbles that must have been carried south by icebergs and dropped into the sea as the icebergs melted. This has prompted scientists to visualize "armadas of ice" drifting out of the north as the last glaciation began, dumping loads of sand and gravel onto the seafloor. This interval records a rapid change to very cold temperatures, punctuated by great swings between warm and cold lasting as little as ten years. Then the cold came to stay. These fluctuations have caused scientists to warn that our present stable climate may be more unsteady than we assume, and could be tipped very quickly into extremes that would be disastrous for civilization.

The burning of fossil fuel, of course, is the wild card in calculating future climate. Will our own folly actually save us from the coming ice?

Human-induced sources of gases such as carbon dioxide, methane, chlorofluorocarbons, sulfur dioxide, and nitrogen oxides are trapping the Sun's heat. Right now, glaciers are in retreat in many places in the world, from the Cascade Mountains to parts of Antarctica. With scientists predicting that atmospheric carbon dioxide will double over the next century, a new computer model developed by scientists from the University of East Anglia in England has factored in man-made global warming and predicted that this could delay the next ice advance by perhaps as much as fifty thousand years. However, when the ice does return, it will be an even more extreme glaciation than otherwise might have occurred, according to their calculations.

Other scientists fear global warming may actually *trigger* the next glacial advance. The reasoning behind this paradox comes from new understandings of atmospheric and oceanic circulation. The atmosphere transfers moist air evaporated from the tropics toward the poles, and the warmer the air, the bigger the transfer. There is the strong possibility that increased moisture in the far north could cause more snowfall and start the expansion of glaciers in that region, even though the overall temperature of the Earth was undergoing a short-term rise. Moreover, the short-term warming would also coax a melting of ice during the summer months, increasing the amount of freshwater entering the ocean circulation systems. This could disrupt ocean currents and affect weather in highly unpredictable ways.

Whereas most of the populous parts of North America lie at latitudes between about 30 and 45 degrees North, most of the population of Europe is about ten degrees farther north: London and Paris are nearly at 50 degrees N, Berlin at 52 degrees N,

Copenhagen and Moscow at 56 degrees N, and the cities in Scandinavia at 60 degrees N. Yet in spite of this, the European subcontinent is extremely productive in agriculture. It supports twice the human population of North America on a much smaller landmass, because its warmth comes from the Gulf Stream. This current, originating in the Gulf of Mexico and Caribbean, flows up the Eastern Seaboard of North America and then vaults across the Atlantic to deliver warmer water to northern Europe, keeping it 5 to 10 degrees Centigrade warmer than it otherwise would be.

One branch of the Gulf Stream carries warmer water to the vicinity of Iceland and Norway. Eventually it cools, sinks deeper into the ocean, and returns south as a cold deepwater current. As it sinks it carries more salt, for saltier water is heavier and tends to sink because of its great density. Warm, fresher water thus travels north on the surface and returns south as cool, saltier water. Paradoxically, this system would be shut down if even more freshwater were added to it on the sea surface from glacial melting caused by global melting: The Gulf Stream would begin its deepwater return south before it traveled so far north.

The failure of one single current would, at first glance, not seem to be the stuff of sudden global climate change. But the world's oceans are but a single body of water, and heat flow is global. If the North Atlantic current with its warm water and the returning deep cold water system fails, the entire world would experience sudden climate change. Europe, with its 650 million people, would go into a deep freeze. Its ability to feed itself would disappear.

University of Washington scientist William Calvin, who has written about this potential for sudden climate change, describes this scenario as follows:

Plummeting crop yields would cause some powerful countries to try to take over their neighbors or distant lands—if only because their armies, unpaid and lacking food, would go marauding, both at home and across the borders. The better organized countries would attempt to use their armies, before they fell apart entirely, to take over countries with significant remaining resources, driving out or starving their inhabitants if not using modern weapons to accomplish the same end: eliminating competitors for the remaining food. This would be a world-wide problem—and could lead to a Third World War.

Calvin calculates that Europe would have a climate like that of present-day Canada, and instead of being self-sufficient in food as it is today, it would be able to feed only one out of twenty-three inhabitants.

What makes sudden global cooling especially catastrophic is that it does not affect a relatively small area such as a hurricane, tornado, or earthquake. Nor is it of short duration. Calvin argues that even an abrupt meteor strike that killed a majority of the human population in a short period of time would not be as disastrous in the long term as a new global glaciation. And unlike the end of the last glaciation, when there may have been at most 2 million to 3 million humans scattered around the globe and needing to feed themselves, today there are more than 6 billion. Human population is expected to exceed 10 billion by 2050 to 2100, assuming an annual increase of 1.6 percent. While this rate is somewhat reduced from the 2.1 percent characterizing the 1960s, it remains a staggering figure. By 2150, according to United Nations estimates, population *could* climb as high as 28 billion, or nearly five times as great as today. Is this possible to sustain?

In his 1995 book *How Many People Can the Earth Support?* author Joel Cohen suggested that "the possibility must be considered seriously that the Earth has reached, or will reach within half a century, the maximum number the Earth can support in modes of life that we and our children and their children will choose to want. . . ."

And of course these estimates are for the planet as we know it *now*, not the planet of a coming Ice Age glaciation. As the next ice advance begins, the world's climate will become cooler and drier, and the broad geographic belts of grasslands (which now supply so much of humanity's food) will contract in size. Food production will be compressed into narrow zones at midlatitudes, squeezed between an ice cap to the north and deserts to the south. Half the world's population would have to move to lower latitudes. The developed nations would demand living space from their southern neighbors, and when those demands were refused, war could occur. Every parent will do what is necessary to feed his or her children.

We humans are blinded by the moment we live in, the brief ten thousand years of aberrant calm and warmth that marks this present interglacial. The reality is that such moments are rare and quickly pass, to be replaced by, on average, ninety thousand years of numbing cold, ice, dust, and drought. Enjoy this summer. The forecast is for a long, brutal, and seemingly never-ending winter.

But end it will, in perhaps 2 million to 10 million years from now, as the continents drift southward and the landmass available for glacial ice caps recedes. Then the time of ice ends forever. Eventually there will come a time, 250 million years from now, when there will be but a single ocean on the planet, and a single continent. It will be the time of a new geography. Will humans be part of that future?

· 5 ·

# THE RETURN OF
# THE SUPERCONTINENT

IN THE LAST DAYS OF THE PALEOZOIC ERA, SOME 250 MILLION
years before the emergence of *Homo sapiens,* a small cynodont
trotted across a parched landscape in the vast interior of the southern continent we today call Gondwanaland. It rapidly skirted the
copses of dying glossopterid trees and low gingko shrubs to get to
the nearest water hole. Rotting skeletons of newly dead animals lay
about this bleak oasis, hardly scavenged, simply dissolving into
bacteria-rich slime under the immensely powerful Sun. The temperature was normal for this early morning time of day, over
40 degrees C (104 degrees F), and the cynodont could not bear to
spend much time out of its burrow. A quick drink, and then back
to the safety and coolness of its deep underground lair, where its
mate suckled a new litter. The cynodont was a small predator only
a foot and half in length, and looked something like a future evolutionary descendant that it would someday spawn: a small dog. But
although roughly doglike in size and shape, the cynodont would
never have been mistaken for the cute doggie in the window were it
resurrected in the time of man: it was a dog without hair, a dog

with scaly lizardlike skin, cruel downward-pointing fangs, yellow-slitted lizard eyes, and a brain barely larger than that of a pigeon. It was bent on survival in a world gone hellish. It was not a mammal—yet—but it would be the seed from which all mammals would spring. It was also soon to belong to one of the lucky few on Earth: its species would survive the most hideous mass extinction ever, a period of mass death that killed off nearly 90 percent of all organisms on the planet.

The cynodont spent the nighttime hours seeking insects and the daylight hours deep underground, sleeping. But finding food and water was becoming increasingly difficult, even here near the South Pole, where the temperatures were still—barely—cool enough to allow animal life. Farther north, nearer the equator, the Earth was already sterilized of all animal life save for insects and spiders. There was not much life to be found anywhere on the planet, and each year it dwindled.

The cynodont approached the water hole warily. It barely cast a shadow even though the day was cloudless, for the sky was hazed with volcanic dust and immense volumes of carbon dioxide and methane, causing the planet beneath to broil like the interior of a greenhouse on a hot summer day. As usual a few of the piglike lystrosaurs were grazing stolidly around the shrinking pond, skirting the large mud cracks to feed upon a few still-living reeds and the tubers that gave rise to them. The cynodont was the most intelligent creature ever to have evolved on Earth until that time, intelligent enough to be wary of whatever might hide in the piles of large boulders adjoining the water hole. It began to lap up the hot, fetid water, and then froze as a flash of movement caught its eye. Past the cringing cynodont the menacing shape of a gorgon lunged, ten feet of coiled muscular fury. Here was the largest predator that the vertebrates had yet produced, the nightmarish top carnivore of the late Paleozoic world. The gorgon bowled into the now-squealing lystrosaurs, slashing into one terrified animal with its giant saber teeth, its great jaws disemboweling the herbivore. But the rush of the gorgon had carried both it and its prey into the center of the pond, a region where black oozing tar was barely concealed by a thin layer of fresher water. All four legs of the gorgon became mired in the black ooze, and slowly its predicament became apparent. The huge reptilian head raised skyward and roared in fury, but it could not break free. The cynodont would hear the roaring for several days from the safety of its burrow, and then no more.

The gorgon's skeleton eventually sank into the ooze and was covered by sediment. The bones held together in the sticky mud even as the flesh rotted away, and as years passed into decades, then into millennia, then into intervals of time counted in the hundreds of millions, its lithified remains became deeply buried under thousands of feet of sedimentary rock. Far above its resting place the first dinosaurs evolved, small and timid, and then increasing in size and number until they ruled the Earth. The huge supercontinent of Gondwanaland split asunder, its pieces wandering in various directions across the Earth's surface. The gorgon's resting place became Africa, and more time passed. Mountains rose and fell, and the dinosaurs' long summer came to a shocking end in a single day in the form of an impacting asteroid. Tiny mammalian survivors—all descendants of the cynodonts—inherited the Earth; they multiplied, diversified, and one branch of them left a life in trees to become erect, ground-dwelling bipeds with giant brains. These were creatures with curiosity, and they dug up the bones of the long dead simply to satisfy that curiosity.

Accordingly, 250 million years after its death in a small drying pond on an aging planet, the long slumbering bones of the gorgon were disturbed for the first time. A hammering of blows rained down on the quiet crypt, and the Sun shone once again on the gorgon. The empty eyes of the gorgon's skull faced skyward, and had they been able to see they would have beheld the face of Peter Ward.

Peter was on the hunt of the greatest mass extinction in our planet's history, an extinction greater than that of the dinosaurs, and greater than that of the Ice Age. He has marched backward in time to understand the next great catastrophe that might be in store for planet Earth.

Our data and models predict that just as the world's continents once formed a supercontinent we call Gondwanaland a quarter billion years ago, they may drift together to form a successor to Gondwanaland a quarter billion years in Earth's future. And once again, it may cause a mass extinction that kills off the majority of species on the planet.

IMAGINE WE ARE CATAPULTED FORWARD IN TIME TO THIS UNIfied landmass. We would encounter a world as strange as that of glaciers of the Ice Age. Our first observation is that the air stinks, like swamp gas, and the humidity is nearly 100 percent. The temperature is well over 38 degrees C (100 degrees F). We are at the edge of the sea and the source of the odor becomes apparent: we can see that the sediment is nearly black, burping bubbles of methane. There is no intertidal life, no barnacles or mussels. The sea seems dead. Then we notice some strange mounds dotting the bottom and pull one out to look at, releasing a gout of stinking gas. It is stony, layered, and covered with gelatinous goo. It's our old friend the stromatolite, the type of layered bacterial mat that was so common in the time before animals, and which had reappeared during times of mass extinction in the deep past. Here it is again, in our distant future, because the creatures that once would have eaten it are gone.

It smells.

We look out to sea. No jumping fish. No diving seabirds. Nothing. A dead sea, but for the stromatolites.

We turn. The land is richer in life than the sea, with plants and insects in profusion, but it seems somewhat impoverished when we look in detail. Even though we're in a very hot place, a place that in our world would be crowded with biodiversity, we soon notice that the vegetation is made up of very few species, repeated endlessly. It's as if the simplicity of an Arctic ecosystem, made up of a few species, has been transferred to the tropics. We look at the Sun. It seems brighter. Is this our imagination? It is certainly hot enough. This place is like the worst of the Midwest on a hot day in summer, but how can the Midwest climate be found here at the seashore?

Is this the end of the world? No. Is it a dying world? Most assuredly. Peter feels he has been here before because this world seems to be a double of the 250-million-year-old Permian world—when the gorgons ruled, and the cynodonts survived.

What caused that greatest of mass extinctions? It is one of the greatest mysteries of science. Was it caused by an impact of an asteroid or comet? While some Earth scientists think so, there is still not conclusive evidence that that is the case. A prevailing view is that planet Earth did it to itself by clumping its continents together, and our mathematical models of continental drift foresee a return to the same conditions. Perhaps the next such event will end all animal life, instead of nearly doing so. What could cause this calamity? Perhaps the very fecundity of this bizarre futuristic world might be a cause. In the past there have been times when the accumulation of rich organic sediments have created short-term biological crises. Another such crisis may face our planet 250 million years from now. Will it kill off all life? Doubtful. Will it change the course of evolution? If the past is any guide, that's a virtual certainty.

The key to our prophecy is the future positions of continents. Continental drift is going to move enough of Eurasia and North America out of high latitude, and enough of Antarctica out of the extreme southern latitudes, to eventually end the tyranny of ice. We have a good idea of how and even when that will come about, based on powerful new computer simulations of plate movements. Plate movements for the past 600 million years are roughly known, and the positions of continents during the past 200 million years is very well known. When complex animals first appeared during the

Cambrian Explosion, the continents were widely dispersed along the equator. For the next 200 million years large-scale drift and continental collision resulted in the formation of ever-larger land bodies as well as major mountain chains, including the Appalachians of the eastern United States. By about 300 million years ago the major continents had coalesced into a single united block, a "supercontinent" we call Pangaea, and 50 million years later the huge southern continent of Gondwanaland had formed. Then it too broke, eventually leading to the map we see today. North America split from Europe and South America from Africa, creating the Atlantic Ocean. By 120 million years ago, the southern continents broke apart as well, with Africa, Antarctica, India, and Australia moving in divergent directions and culminating, today, in the continental positions so familiar to us.

Continental drift is continuing: the Atlantic Ocean, for example, widens every year at about the rate your fingernails grow, and the Pacific shrinks by the same amount. The convective cells of the mantle continue to boil, plate tectonics continues to operate, and continents continue to float about.

Yet scientist Chris Scotese, whose Paleomap Project has mapped drift in the past, believes this trend will end after several million more years, and then the trend will reverse. "Two hundred fifty million years in the future," he projects, "the Atlantic and Indian Oceans have been closed. North America has collided with Africa, but in a more southerly position than when it rifted. South America is wrapped around the southern tip of Africa, with Patagonia in contact with Indonesia, enclosing a remnant of the Indian Ocean. Antarctica is once again at the South Pole and the Pacific has grown wider, circling half the Earth."

Past continental positions are determined by using a technique known as paleomagnetism. Ancient rocks, when solidifying, can freeze into their composition as an indication of the latitude at which they were formed, by preserving information imparted by the Earth's magnetic field. By sampling thousands of such data points, extremely detailed maps of past continental configurations have been assembled. Establishing future continental positions is done by looking at present-day movements and positions and then extrapolating forward.

This future motion will have enormous effects on future climate and on the fate of future life—and on the very nature of the Earth systems themselves. By fifty million years from now the Mediterranean Sea will be gone, its space taken by an enormous mountain range extending from what is now Europe to the Persian Gulf. Australia has moved northward, closing with Papua New Guinea and Indonesia, while Baja California has slid northward along the Pacific Coast of North America. But far more important than these new continental positions will be the formation of new subduction zones, the regions where the Earth's crust dives beneath the continents. Today we know that subduction is initiating in the central Indian Ocean and in the ocean off Puerto Rico. These events suggest that new subduction zones will be in place off both eastern North and South America. As this happens, mountain building will be initiated once again in the Appalachian regions, and along the eastern coastline of South America as well. These regions will become home to gigantic active volcanoes and rising mountain chains.

It is not just the positions of the mountains that will change. As Antarctica drifts northward its vast ice sheets will melt, and warming temperatures will melt Greenland as well. The oceans will rise nearly three hundred feet higher than they are today, and then still farther as new midocean volcanic ridges swell, displacing water, and spill oceans farther onto low-lying land surfaces. The Amazon Basin, the Gulf Coast plain of North America, and the western portion of central Africa will all flood. Antarctica will be split in half as a new spreading center cleaves it in two. All coastal plains will flood; all modern-day deltas will be lost. If not an end of the world, this rising of the sea will spell the end of geography as we know it.

The flooding of the continental margins by the rising sea will cause a radical climate change. There is a strong possibility that Earth will again experience the sort of climate that was present long ago during the Mesozoic era, and especially the Cretaceous period of about 120 million to 65 million years ago. At that time, near the end of the Age of Dinosaurs, the planet was a humid garden of high heat, with very little temperature difference from the equator to nearly the poles. In such a world the circulation pattern found in today's oceans will radically change. The oceans, as they had been for long intervals during the Paleozoic and Mesozoic eras, will return to an unmixed state. That means they will again become stratified and, to a degree, dead. When this happens the flow of atoms in the carbon cycle—one of the systems that we have stressed as being key to habitability and as being analogous to the organ systems of a living individual—will radically change. This change may also affect the rock circulation systems, creating new tectonic conditions that are unlike any currently present on earth—including the possibility of "flood volcanism" in which huge volumes of magma spew forth onto land surfaces. These episodes in past times are strongly linked to profound global mass extinction of life.

The change of climate in this far future will first affect the circulation patterns of the oceans. Like our recycled rocks of an earlier chapter, oceanic circulation patterns are analogous to the bloodstream of an animal. In organisms the circulation system is designed to bring oxygen and nutrients to all parts of the body. In oceans, it redistributes heat. Ocean currents take energy from the hot equatorial regions and export it to the cool poles. Then they dive to deep water and transport cold polar water toward the equator. This circulation is vital to keeping climate moderate on much of our planet. It also carries oxygen from the top to the bottom of the sea. Because of this, ocean water has roughly the same amount of oxygen in its depths as near the air. During the past, however, when global temperatures were higher, oceanic circulation systems were either sluggish or nonexistent. For enormous amounts of time the bottoms of the oceans held less oxygen than the surface. As a consequence great areas of ocean bottoms became much like the bottom of the Black Sea today—anoxic, or without oxygen for fish and other marine creatures to breathe.

The warm Cretaceous climate produced this condition. There is abundant fossil and geochemical evidence that during the reign of the dinosaurs, warm-water marine animals and plants fled almost to the poles. Globally warm seas prevented the sinking of cold water at high latitudes and so the depths were never replenished with oxygen. As sea levels rose, the deep, anoxic waters spread over the edges of the continents, producing a sticky, stinking goo that became black shale deposits. Moreover, by shutting off the normal circulation of the oceans that involves surface-water sinking and deepwater upwelling, the deep ocean can build up huge amounts of carbon dioxide, and carbon in the absence of oxygen. This material can become an underwater time bomb or, better yet, an underwater chemical weapon against the biosphere—if it is rapidly released.

There is a modern-day analog to this future stagnation: the Black Sea. It is anoxic below because so much freshwater pours into it from large rivers such as the Danube, Dniepr, and Don that its surface water is too light to sink: it doesn't have enough salt. It is the world's largest stable anoxic basin. Black Sea waters deeper than 100 meters are characterized by an absence of oxygen and elevated concentrations of hydrogen sulfide and methane. If the world's oceans as a whole were to lack vertical circulation like the Black Sea, more than 90 percent of the water on Earth would be anoxic—just as it was in the Mesozoic era. And this is the state of the oceans we predict will take place 250 million years from now, as continental drift changes both climate and geography to a state uncannily similar to those of the Mesozoic. We cannot predict a return of the dinosaurs, but we can predict a return to a worldwide anoxic ocean—an ocean that would seem totally alien to our mixed-ocean world. Not only would such oceans be more lifeless than our own, but the accumulating greenhouse gases in their depths could suddenly erupt, causing sudden global warming.

THE PATTERN OF SMALLER CONTINENTS ASSEMBLING INTO A supercontinent and then breaking apart again has been dubbed the Wilson Cycle, in honor of one of the pioneering discoverers of plate tectonics, J. Tuzo Wilson. The entire cycle seems to take about 500 million years, and there is no reason to believe that plate movement in the future will alter this trend. By about 100 million years from now, the continents will have reached their maximum separation and begin to coalesce. In 150 million years the Atlantic will have become far smaller, and the Pacific Ocean will increase in size as the continents all begin a mad rush toward mutual collision. By 250 million years, Europe, Africa, North and South America, and Asia will have fused into a single continent while Antarctica and Australia will remain aloof. Because of the presence of subduction zones virtually encircling this supercontinent, a wall of mountains will enclose much of the land surface, walling off its interior. Just as the Great Basin of the western United States is walled off by the Rockies to the east and the Cascades and Sierras to the west, the whole of the new supercontinent will be encircled by mountains.

This will not just change the world map. It could radically change the nature of life on Earth, because the interior of this new supercontinent will have a far harsher climate than today's continents, which have weather modified by the ocean. The resulting extremes of heat and cold and severe drought, caused because marine storms are blocked by the ring of mountains, will mean either a major mass extinction of life, or—less probably—the *complete* extinction of animals and plants on the planet.

We predict this because of the world of the cynodont and gorgon. There have been five major mass extinctions in the past 500 million years of the Earth's history, each eliminating more than half of all the animal species on the planet, but by far the greatest was

255 million years ago

Today

250 million years from now

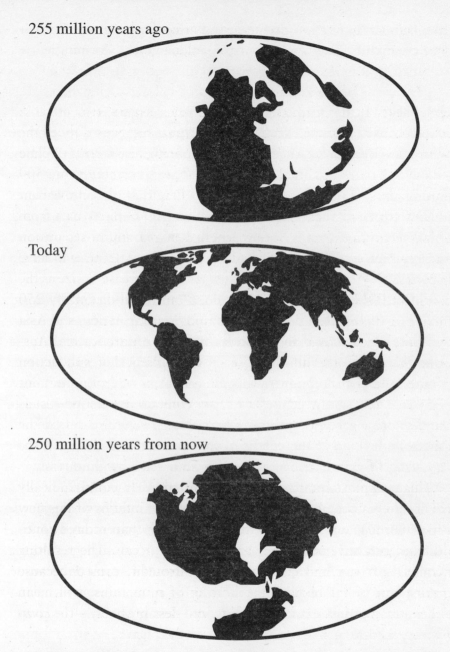

Most of the world's landmass was assembled in a single great continent 255 million years ago. After spreading to the world we know today the continents are predicted to merge once again 250 million years from now. Figure derived from Chris Scotese's Paleomap Project.

the Permian extinction that occurred at the last great assembly of the continents, near the end of the Paleozoic era. Some scientists estimate that more than 90 percent of all species disappeared.

The cause of the Permian extinction has long been debated, but one theory is that formation of the Pangaea supercontinent caused rapid climate change. With a larger continental interior, climate extremes were pronounced. Climate simulations for the Permian show that the continental interior would have had summer temperatures above 38 degrees C (100 degrees F) and winter temperatures below freezing. Such extremes have significant effects on limiting biodiversity, as anyone visiting America's Great Basin can observe. But is such extreme climate enough to kill off nine out of ten species?

Most Earth scientists now consider that several events combined to create this greatest of mass extinctions. Perhaps a comet or asteroid hit, but the Permian extinction was more protracted than the one that killed the dinosaurs, so it is unlikely this was the sole cause. Scientists do know there were eruptions of enormous floods of volcanic basalt, which released vast volumes of poisonous sulfur and chlorine gas into the atmosphere. And the carbon dioxide and methane lurking in the depths of the stagnant oceans could have suddenly erupted, either poisoning animals outright, causing a methane explosion, or triggering runaway global warming.

There is a contemporary example of what could have happened. In 1986, a giant cloud of water vapor and concentrated carbon dioxide belched out of Lake Nyos, which fills an ancient volcanic cone lake in the African equatorial country of Cameroon. A white mist began bubbling out of the lake, followed by an explosive burst of water creating a wave two hundred feet high. This cloud and wave raced ashore like a ghostly dense fog, traveling twelve miles before it dissipated. At least seventeen hundred humans died from Hypercapnia, the medical term for breathing too much carbon dioxide. So did all the dogs, birds, pigs, and cows in villages near the lake. Imagine such an eruption on a global scale!

Northwestern University geologist Gregory Ryskin has done just that, and his scenario is even more catastrophic. A stagnant ocean could hold and erupt methane with an explosive power twenty thousand times greater than the world's stockpile of nuclear arms, he calculates, and a mix of air and methane could spread across the planet and ignite a firestorm. It would burn all vegetation first, and then create such a thick cloud of smoke that it blocks out sunlight and freezes any wretched survivors. And then, in a final blow, "after the ash and dust have settled, carbon dioxide, together with remaining methane, create the greenhouse effect which may lead to global warming. . . ."

Whatever the exact cause, the prospect of a future supercontinent is grounds for pause. Will that future mass extinction occur with sufficient ferocity to kill off some appreciable fraction on Earth's cargo of life? We can't predict that with certainty. Still, computer models suggest that by 250 million years from now, the level of global productivity will be far lower than in the present day. We will have traveled half a billion years from one Permian extinction to another. And yet even this future world bears some familiarity. As we look further into the future, a far stranger fate awaits our planet—the end of almost all plant life on Earth.

· 6 ·

# The End of
# Plant Life

AT SEVERAL POINTS IN THIS NARRATIVE, WE HAVE MENTIONED
that the Sun has been getting hotter: that its brightness, in fact, is
30 percent greater now than when the Earth was formed. If this
information has left you uneasy, your instincts are correct: our
solar companion is going to cause increasing problems for life in
the future. And if you are wondering why a nuclear furnace with a
set amount of fuel is getting hotter the longer it burns—instead
of cooler, like an unnourished campfire—then you've got a good,
commonsense question.

The answer lies in the physics of solar combustion, and we must
take a moment to explain this because it is critical for what is
to come. Paradoxically, the more hydrogen the Sun converts to
helium, shedding energy in the fusion process, the hotter it is going
to get until its final, fiery expansion.

The Sun is conceptually simple: it is just a big ball of hydrogen
gas with a nuclear reactor at its center. It is a star similar to others
in the sky, though obviously of more importance to us since we
orbit it. The Sun's gravity holds us in orbit, its heat keeps us from

freezing, and its light powers photosynthesis, the conversion of sunlight to food sugars by plants, which is the very foundation of our food chain. The Sun radiates 60 million watts of power from every square meter of its surface, and even at Earth's distance of 150 million kilometers, the light of the midday summer Sun carries a thousand watts of power to each square meter of our planet. So bright is the Sun that daytime is a million times brighter than the night.

Although this output is quite impressive, a star like the Sun is actually very simple to make. It requires only the most basic of materials and essentially no assembly instructions. In fact, our Milky Way galaxy makes dozens of new stars each year. The recipe for making our Sun is as follows:

1) Take a big enough batch of hydrogen gas.
2) Let it collapse under its own gravity to form a star.

The trick is to have enough mass, the Sun's only real secret for success. To make the Sun you need nearly a million times as much mass, or material, as makes up the Earth. When such a large mass of hydrogen (plus minor amounts of helium and trace levels of all the other elements) contracts under its own gravity, its interior heats to temperatures of millions of degrees and its hydrogen atoms begin to fuse into helium, a reaction that gives off heat energy. At the end of the gravitational collapse that gives it birth, the Sun reaches a stable configuration where the weight of its upper layers is exactly supported by the pressure of its hot interior. To understand this, imagine that the core of the Sun is enclosed by a spherical balloon. The pressure inside the balloon would be just sufficient to support the weight of the matter above, and so the balloon neither expands nor contracts.

Remarkably, in stars like the Sun, the stable balance between gravity can be sustained for billions of years. The vast amount of heat lost to space is precisely replaced by energy generated in its interior. The source of this energy is thermonuclear fusion, or the fusion of hydrogen to form helium. Helium, the product of this fusion, is slightly less massive than the materials from which it forms, and the missing mass is converted to energy following Einstein's famous equation $E=mc^2$. This conversion of mass to energy is the powerhouse that runs the Sun. While 4 million tons of mass is totally converted to energy each second in the center of the Sun, our star is so immense that it can generate its energy for 10 billion years before it runs out of hydrogen fuel.

Energy production in the Sun occurs in its core, where the density of matter is a hundred times higher than that of water and the temperature is 15 million degrees Centigrade. The first critical step in the chain of reactions that lead to helium is the fusion of two protons (normal hydrogen) to form deuterium, the heavy isotope of hydrogen composed of a proton and a neutron. The simple fusion process is extraordinary: not because it is vigorous and violent, like a hydrogen bomb, but because it is so slow. In the conditions that exist in the center of the Sun, a single proton will collide with other protons more than a trillion times each second. Yet when two protons collide, their identical electrical charges almost always cause them to repel each other and they simply bounce off. To form deuterium, they must hit hard enough to fuse, and the hotter the gas, the harder the protons collide with each other. Yet even at 15 million degrees C, the probability of fusion is almost zero. It takes a typical proton 10 billion years to have even a 50 percent chance of actually fusing with another proton—this after colliding a trillion times a second for 10 billion years!

This abysmally slow process is the fundamental reason why the Sun lasts so long and Earth's life has had time to evolve. From the viewpoint of an Earth-based consumer, the Sun is a fabulous source of endless energy: predictable, reliable, and free. But from a power engineer's perspective the Sun is a woefully feeble power plant. As hot as the Sun is, it takes half a million tons of Sun to produce a single kilowatt of power: not enough energy to run a hair dryer. To put this into perspective, three average people generate about a kilowatt of heat just from the chemical energy of eating breakfast, lunch, and dinner. Comparatively, the Sun is a feeble power plant but this is one of its most important attributes. With its great mass it can generate vast amounts of power for billions of years. Again, the secret of the Sun's origin, output, and longevity is simply its enormous mass. It is not just the complexity of nature that is so fascinating, but its underlying simplicity. Fundamentally, the Universe runs on the mere gravitational lumping together of its simplest atom, hydrogen. From this comes energy and, ultimately, the more complex elements needed to build planets and us.

We joke about expecting the Sun to come up in the morning, but just how stable is it? We can actually judge this by observing similar stars. Stars that have a similar mass and age as the sun also have nearly the same brightness as the sun. This suggests that the brightness of these Sun-like stars does not vary greatly, given the same age. But it does change over time. Very slowly, our Sun is getting brighter.

Paradoxically, the Sun is becoming brighter because the number of atoms in its center—its fuel—is decreasing. How can this be? Think back to the imaginary balloon enclosing the core of the Sun. The pressure on the inside has to support exactly the weight of the overlying mass. We know that the size of the Sun does not change over long periods of time, so the pressure in its center must remain reasonably constant. The pressure is produced by the cumulative impacts of vast numbers of particles. As each atom bounces off the surface of the imaginary balloon, it imparts a small outward force and the total pressure is the net effect of all of the particles in the balloon. Using what is known as the "gas law," the pressure in the balloon (of constant volume) depends on just two things, the number of particles in the balloon and the temperature of the gas. The hotter the gas, the faster the particles fly. Since the number of particles is constantly decreasing as the Sun burns its fuel—if all the hydrogen was converted to helium, the Sun would have only one-fourth the number of atoms it started with—its temperature must constantly rise if the pressure is to remain constant.

As the temperature rises, hydrogen travels at higher speed, collisions are more energetic, more protons fuse, and the production of helium rises. So does the total amount of released energy. As crazy as it seems, the gas law requires that the more hydrogen the Sun burns, the hotter it has to be to avoid collapse. This slow ramp-up of energy generation occurs over the full 10 billion years of the Sun's lifetime. Eventually the Sun will have consumed most of the hydrogen in its center and the remaining hydrogen shell, around a helium-rich core, will swell until the Sun becomes a red giant and its brightness increases thousands of times. This fate comes late in our story—but, already, the Sun has increased in brightness by about 30 percent in the past 4 billion years.

We know that for all of its history, the Earth has had the good fortune of being within the "temperate zone" of the solar system that allows liquid water to exist on its surface. This habitable zone extends to a limit just inside Earth's present orbit—about 10 million miles away, toward the Sun—to a less-understood outer limit near Mars or possibly beyond. As the sun brightens, the habitable zone will move outward, effectively bypassing Earth's orbit in 500 million to 1 billion years. The consequences of this will be explained shortly.

Why didn't the Earth freeze early in its history when the Sun was dimmer? Because the atmosphere had higher concentrations of greenhouse gases such as carbon dioxide to capture the relatively feeble heat. Why doesn't our planet turn into an oven like Venus now, as the Sun grows brighter? Because of the subsequent sharp drop of carbon dioxide, the regulation and sequestration of which was explained in earlier chapters. Earth has managed to sustain a cool enough temperature to support life through the storage of high amounts of carbon dioxide in such "reservoirs" as forests, soil, oceans, and rocks.

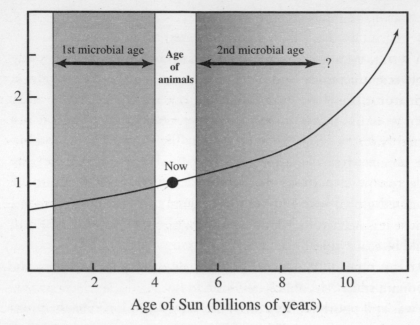

Age of Sun (billions of years)

For its first 10 billion years, the Sun slowly increases in brightness. The effect is gradual but it is the major reason why Earth's biological capabilities change dramatically with time. The brief age when evolution and environmental conditions allow plants and animals to exist is bracketed by two long-duration ages when the Earth is solely inhabited by microbes invisible to the naked eye.

Yet this is a game that can go on for only so long. Carbon dioxide is already only a trace gas in our atmosphere. As our planet continues naturally to sequester it to regulate its temperature, primarily by silicate weathering, it will lose the carbon dioxide that is necessary to sustain plant photosynthesis, the energy base of almost all life and the primary source of free, breathable oxygen. For billions of years our planet has maintained a careful biological balance. Some 500 million to 700 million years into the future, the world will turn brown.

. . .

JUST WHEN THIS WILL HAPPEN HAS BEEN A SUBJECT OF CONSID-
erable scientific study and debate. It began with James Lovelock,
originator of the Gaia hypothesis that our planet is literally alive.
When would it die? In a pioneering paper published in the science
journal *Nature* in the 1970s, Lovelock and coauthor Mike Whitfield pondered this question. They presciently pointed out that
while too much carbon dioxide is a bad thing that can cause
unwanted global warming, too little would be equally disastrous,
because it is needed for plant growth. Without plants, life on Earth
would be scant indeed.

At the time of the pioneering Lovelock-Whitfield article, the
carbonate-silicate feedback system had been only newly proposed
and was still poorly known and little accepted. Nevertheless it was
clear to Lovelock and Whitfield that, in the future, as the Sun
became brighter and the increased solar luminosity gradually
warmed the Earth, silicate rocks should weather more readily,
because warmer temperatures cause more wind, rain, and erosion.
This would cause atmospheric $CO_2$ to decrease. The genius of their
work was in comprehending that there would come a time in the
future when carbon dioxide levels would fall below the concentration required for photosynthesis by plants. Most plants require air
to have at least 150 parts of carbon dioxide for every million parts
of air. Present-day $CO_2$ levels are about 350 parts per million (ppm)
and are rising rapidly due to human causes. Using computer-based
modeling, Lovelock and Whitfield estimated that the end of plant
life as we know it would occur in about 100 million years, because
carbon dioxide levels would drop below 150 ppm. While that
seems very long from now, 100 million years is, in reality, a very
short time for a planet that has had life for at least 3,500 million
years, and microscopic plants of some sort for more than 2,000
million years. It would mean the Age of Plants is already 95 percent
completed! This result came as quite a shock.

With the publication of the pioneering Lovelock and Whitfield paper, the idea that sophisticated models could be used to forecast future events on the Earth was taken up by a succession of preeminent scientists. One such group, headed by Ken Caldeira and James Kasting of Penn State University, increased the sophistication of their model and in their 1992 publication titled "The Life Span of the Biosphere Revisited," published in *Nature*, Caldeira and Kasting improved the models of Lovelock and Whitfield and came up with a more reassuring future.

They pointed out that the Lovelock-Whitfield assumption that plant life requires a minimum of 150 ppm of atmospheric $CO_2$ isn't strictly true. While this is the case for the vast majority of plant species on Earth today, there is a second large group of plants, including many of the grassy species so common in the midlatitudes of the planet, that use a quite different form of photosynthesis that can exist at $CO_2$ concentrations as low as 10 ppm. These plants would last far longer than their more carbon dioxide–addicted cousins, and would considerably extend the life of the biosphere.

With the new calculations and values included, Caldeira and Kasting concluded that the critical 150 ppm value of $CO_2$ would disappear not in the 100 million years in the future predicted by Lovelock and Whitfield but five times that time, or 500 million years into the future. Some plants, using far lower levels of $CO_2$, might exist for as long as another billion years, they added. So, all in all, a rosier picture, or at least a world where roses could exist for another 500 million years.

But Caldeira and Kasting asked not just when plants would disappear, but what amount of life will be present on Earth. They tried to put future numbers on biological productivity, or the rate at which inorganic carbon is transformed into biological carbon through the formation of living cells and proteins. Here their results were rather shocking: from the present time onward, productivity will plummet. Even though life will continue to exist, it will do so in even smaller amounts on the planet—and not a billion years from now, or even a hundred million years from now, but from our time onward! Here is one end of the world, at least as we currently know it: the end of a biosphere as crowded with life as we enjoy and take for granted today.

They also ominously noted that "the problem of the life span of the biosphere has implications not only for the future of our planet, but also for the probability of finding biologically active planets in our galactic neighborhood. . . ."

Planets don't have as much time to flourish as we once assumed. Even if we could find another Earth, we would have to discover it during the relatively brief 10 percent of its existence, or 1 billion years, when it had flourishing plants and animals.

The models used to predict the end of the biosphere continued to be improved, and even better estimates—based on newly recognized rates of weathering and $CO_2$ flux—continued to be published. In 1999, Siegfried Franck and two colleagues improved on the Caldeira and Kasting model, looking backward as well as forward. Their results suggest that photosynthesis will end between 500 million and 800 million years in the future and that about a billion years from now the temperature of the Earth will rapidly rise to unbearable values.

This paper was by no means the last word. Other articles with slight refinements have appeared since, but there seems to be a convergence on a time, somewhere between 500 million and a billion years from now, when land life as we know it will end on Earth, due to a combination of $CO_2$ starvation and increasing heat. Moreover, global productivity is already in decline, and has perhaps been declining for the past 300 million years, as carbon dioxide levels have dropped as plants struggle to maintain a habitable temperature. Life has played a clever game to keep the Earth habitable for itself, but it has bought time only at the cost of steadily lowered productivity because life-building carbon has necessarily been sequestered away. In the endgame with our warming Sun, we living organisms can win for only so long.

...

SUPPOSE WE FLY AHEAD FROM OUR NEW SUPERCONTINENT BY the device of a time-lapse movie, each half a million years compressed into a single frame. As the flickering images progress, we witness a plant-covered planet far in the future that, at first glance, seems utterly familiar. There are forests, grasslands, shrub lands, deserts, high chaparral, taiga, and the many floral groupings so familiar in our own world. The details dance as species evolve but on the whole the future looks like business as usual, plants persisting as they have from 400 million years ago to the present.

And then, as our film continues, the plant-covered world becomes surreal.

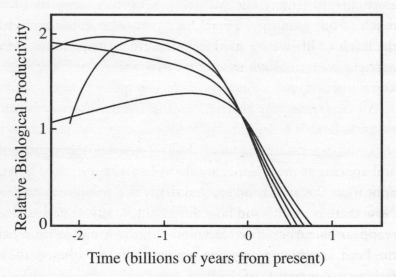

The biological productivity of Earth has increased with time, perhaps already reached its peak, and will decline in the future, in part due to the decline of carbon dioxide. The graph shows predictions based on different models of Siegfried Franck and colleagues. There are several factors that may delay the timing of the drop of productivity to near-zero values but the effect will surely occur as the Sun becomes brighter and the atmosphere no longer has sufficient carbon dioxide to support the growth of plants.

First the forests begin to change, and then all of the floral groups show a sudden and radical changeover in composition. Most of the plant species of the world disappear, to be suddenly replaced by new forms. The rain forests radically change, and while new forests take their place, their appearance is different. Gone completely are the conifers and their ilk, to be replaced by a wide swath of bamboolike trees and palms. Grasslands spread. The tempo of change increases and now our images are herky-jerky with forests, grasslands, and shrub lands coming and going in rapid succession as ranges expand and contract due to climate change.

But even the appearance of the planet at the start of the movie would be shocking to us. It would be like looking at a still vigorous, but obviously aged person—there is no doubt of the ravages of age. And in the case of our Earth, the long-term decline in $CO_2$, the essential nutrient of the planet, will by this time have removed much of our plant life. There are not enough nutrients to support the levels of life we are used to today. By this time in the future the nutrient and circulation systems that maintain the Earth as we know it will have become constricted by aging.

All of this is but prelude, for suddenly whole provinces fall away. It is as if a defoliant is dumped on the Earth and a sudden strike of Agent Orange leaves only the brown of naked hillsides. Soil appears as most plants are stripped away, and then soil too disappears as the winds and erosion strip the Earth down to bare rock. Now there is an ebb and flow, for plants nearly disappear and then reappear as carbon dioxide volume in the atmosphere seesaws at the limit at which plants can conduct photosynthesis, allowing a feeble recolonization of the land. Eventually, however, the effort by land vegetation to take back its ancestral homeland becomes feebler. Finally it stops altogether.

The movie comes to an end. There is the slapping of the end of the film strip against the projector, a staccato beat emphasizing the last image, of a planet without plants. Not only is it a desolate world, we can barely breathe in the oxygen-starved air.

Plants are already evolving to avoid this grim end, and the result can be observed in any pasture or lawn. About 8 million to 10 million years ago, with carbon dioxide levels already declining, grasses developed a way to thrive with less carbon dioxide than their ancestors. They use a compound for photosynthesis that harnesses four carbon atoms instead of three, and thus are called $C_4$ plants instead of $C_3$ plants. While they represent only 5 percent of the world's plant species, they fix 18 percent of the world's organic carbon, and they emerged at a time when the world was shifting from being largely forested and tropical to grassier and cooler. Some scientists argue that these new grasses emerged because plants are already being starved of carbon dioxide.

The formation of new photosynthetic pathways is a sure sign that the long-term reduction in atmospheric carbon dioxide is having a profound effect on the biosphere, and over the next hundreds of millions of years, it should produce a decisive change in global floras. Whereas the majority of vascular plants now on Earth are $C_3$ dicots, there should be an increasing transition toward $C_4$ monocot vegetation. That could mean the gradual disappearance of pine, fir, hardwood, and tropical forests and their replacement with vast grasslands, shrubs, and cacti. Alternately, there are already enormously successful grasslike species that have evolved treelike shapes, such as palms and bamboos, and their descendants may form the forests of the future.

There is also the possibility, of course, that plants will continue to evolve other photosynthetic pathways to compensate for lowering carbon dioxide. In this case we can envision some sort of plant life surviving at minimal $CO_2$ levels. Eventually, however, even these last holdouts will die out because all models suggest that this gas will continue to drop in volume, ultimately arriving at the critical level of 10 ppm. Below that, photosynthesis cannot take place.

It seems ironic that plants will begin to die for no reason that is apparent to our senses. The world will not be a hothouse. (Although it will certainly be hotter than now, it may be no hotter than during the Cretaceous period, some 100 million years ago.) All other aspects of the planet will be seemingly normal. Yet plants will begin to die.

It will not be just land plants that are traumatized by the lowered carbon dioxide. Larger marine plants and perhaps plankton will be similarly affected. Marine communities thus will be strongly affected, since the base of most marine communities is the phytoplankton, the single-celled plant floating in the seas. Yet the disappearance of land plants will also cause a drastic reduction in the biomass of marine plankton, even without accounting for carbon dioxide. Marine phytoplankton relies in part on rotting terrestrial vegetation for phosphate and nitrate, brought into the oceans through river runoff. As land plants diminish, the seas will be starved for nutrients and the volume of plankton will catastrophically decline.

The base of the food chains will disappear, a source of free oxygen will decline, and air, water, and the rate of weathering and erosion will all be affected. The loss of plants will also cause global productivity—a measure of the amount of life on the planet—to plummet. But how much? As catastrophic as the loss of multicellular plants will be, there will still be life, and lots of it. For while terrestrial plants will die off, organisms capable of photosynthesis will not. There will still be great masses of bacteria, such as the cyanobacteria, or blue-green algae, that will continue to thrive, for these hardy single-celled organisms can live at carbon dioxide levels below those necessary to keep multicellular plants alive. Bacteria might increase to fill the niche left by dying plants, and there is an unknown but probably gigantic biomass of bacteria in soil and even solid rock that also fix carbon.

But world productivity will be cut at least in half, and life on planet Earth will become far rarer. No longer will falling leaves create giant volumes of reduced carbon that makes its way into the soil, the sea, and the sedimentary rock record. No longer will coal and oil begin its process of formation. There will no longer be spring plankton blooms, or the marine creatures that graze them and are in turn eaten. As land plants disappear the soil will erode, leaving bare rock. This will, in turn, perturb the hydrological cycle. Giant transfers of carbon between the various land, ocean, and sedimentary record reservoirs will occur.

The disappearance of plants will drastically affect the nature of the planet's surface. As roots disappear, and surface layers become less stable, the very nature of rivers will change. The large, meandering rivers of the modern era date back, at most, to the Silurian period of some 400 million years ago, when land plants first colonized the surface of the planet. It takes the stability provided by roots of plants to maintain the banks of meandering rivers. Without plants, a different kind of river exists: braided rivers or streams of the kind found on desert alluvial fans or in front of glaciers. This was the nature of rivers before the advent of land plants, and will be again when they disappear.

The die-off of plants will also accelerate global warming in two important ways. First, as mentioned above, as plants die off their effects on soil and rock weathering will decrease, causing carbon dioxide gradually to increase in the atmosphere once again. Even small additions of carbon dioxide can have drastic effects in a $CO_2$-starved world; each doubling of $CO_2$ volume causes the atmosphere to heat by a degree. Secondly, and more important, plants themselves cool the Earth by affecting the albedo of the planet. Plant-covered surfaces adsorb solar radiation while rock-covered surfaces reflect it back into the atmosphere where it can cause atmospheric heating. As plants disappear, more and more of the planet will become bare rock surface, for it is plants that largely hold soil in place. The soil will erode from wind and water into the sea, eventually lost as sediment in the deep ocean bottoms.

As global temperatures approach, and then exceed 40 degrees C, or 104 degrees F, the planet will begin to die at the equator, and multicellular life will have to migrate poleward. The continents in the vast middle parts of the world will become largely devoid of surface life due to the high heat. Rainfall patterns will change; vast deserts will grow; all that is green will disappear. The midlatitudes will become the new tropics, and the high-latitude regions will become warm temperate. There will be no snow on the planet except, perhaps, atop the highest mountains; there will be no ice caps. In the sterile tropics, rivers will change morphology from the meandering river systems so familiar on our planet today to braided river systems.

The loss of soils will be no less dramatic. As soils wash or are blown away, they will leave behind bare rock, as mentioned earlier. This will change the very reflectivity of the Earth. Far more light will bounce back into space, affecting the Earth's temperature balance. The atmosphere and its heat transfer and precipitation patterns will be radically changed. Blowing wind will begin to carry the grains of sand created by the action of heat, cold, and running water on the bare rock surfaces. While chemical weathering will lessen due to the loss of soil, this mechanical weathering will build up enormous volumes of blowing sand. The surface of the planet will become a giant series of dune fields.

While this event could signal the final extinction of all plant life on land (and perhaps in the sea as well), it is more likely that a long period of time (perhaps in the hundreds of millions of years) will ensue during which carbon dioxide levels hover at the level causing plant death. As the levels drop, plants die off, reducing weathering. This allows carbon dioxide to again accumulate in the atmosphere, and once again small surviving seeds or rootstocks germinate and, at least for some millennia, persist in low numbers. As plant life again spreads across land surfaces, weathering rates will again increase, carbon dioxide will again be reduced, and plants will again die off.

And what about oxygen? All animals, including us, need it. While some scientists have thought the loss of plants will have little effect on atmospheric oxygen values, new studies suggest just the opposite. The loss of plants will shut off oxygen-producing photosynthesis but will have no effect on the most important oxygen "sink"—the oxidation of dead matter and volcanic gases. It is the latter that will most rapidly deplete the oxygen supply. The recent calculation by David Catling of our astrobiology group at the University of Washington suggests that 15 million years after the death of plants, less than 1 percent of the atmosphere will be oxygen, compared to 21 percent today. The decline of oxygen will not only suffocate animals, it will also diminish Earth's ozone layer, which shields us from the Sun's ultraviolet rays.

What a different planet we will have become! From space, the green hills of Earth will look brown and red. Mars will have gained a larger twin. It will be an end of the world—or at least the world we so take for granted, the world of green plants.

But these are the observable features, the outward appearance of the body. What about the inner workings, the Earth systems of habitability that we have profiled in the preceding pages? They are breaking down by this point in life, failing, or failed, killing the planet. With the loss of sufficient $CO_2$ the planet Earth essentially starves to death. The nutrient cycle has stopped. With the breakdown of this system, the body will wither and die in a finite time—the subject of the next chapter.

· **7** ·

# THE END OF ANIMALS

WITH THE LOSS OR DRASTIC REDUCTION OF PLANT LIFE, THE world of animals will be in dire straits, its time now limited. Animals need oxygen to breathe, and that oxygen comes from the photosynthesis of plants. It was bacterial blue-green algae, plankton, and plants that gave our planet a habitable atmosphere. It will be their decline that takes that habitable atmosphere away.

According to some scientific models, the last animals will be asphyxiated from lack of oxygen within a few million years after the end of plants. But perhaps a few plants will linger, and keep some low level of oxygen in the atmosphere sufficient to allow animals barely to exist. What might that bleak world of animals look like, soon after plants are gone but when there is still enough oxygen to sustain them?

It is a world of rocks, sand, and howling winds, with a piercing Sun in the sky above. The sky is no longer the clear blue that we know today, but more opaque, like a smoggy big-city sky. The soil built by plants is gone, blown to the four corners of a brown world and finally lost to the bottom of the sea. Lichens and a rare moss cover the undersides of sandblasted rocks. The Sahara Desert is now world covering.

But there is movement among the rocks, and closer inspection reveals a host of surviving animals. Armadas of insects, spiders, and scorpions are the most common, but lizards, birds, and even small mammals are managing to hold on as well, so soon after the loss of higher plants. Oxygen levels are dropping, but it will take time for chemical reactions to consume it from the atmosphere. Meanwhile sea life is still abundant, but land life is in the final stages of retreat. We scientists know that this time is coming, and we have a good sense of what it may look like—because we know what happened long ago.

For the past 500 million years the Earth has been dominated by animal life. But, as we have seen, it wasn't always so. There is good evidence that life has existed for at least 3.4 billion years—and possibly for 4 billion years. For most of that time, animals did not exist. We have a sense that the beginning of animal life on our planet was like a dawning spring, but the reality is that the Age of Animals was like a baby born to almost impossibly old parents. The truth is that the Earth, as an abode for life, is in at best late middle age, and probably in old age. The Age of Animals is a last hurrah, the farthest height of the cannonball trajectory of life's history, the apogee of organ and ecosystem complexity. With its end, everything begins a relentless simplicity.

What might it look like? Its mirror—a model for the end—is at the famous Burgess Shale fossil bed at the border of British Columbia and Alberta. To understand our planet's future, let's go there now.

Near the small town of Field, a well-worn path leads up the side of shale-covered Mount Stephens. It starts easily enough, amid the U-shaped valleys carved by the glaciers of the last Ice Age, but then it becomes steeper through a seemingly unending succession of switchbacks cut into the layers of rock that make up the mountain. Hours pass as you climb toward the dawn of animals, the Sun passing from morning promise to the maturity of afternoon, and still you rise, ever higher above the cordillera, each new turn revealing another unbelievable vista. Perhaps this is not the roof of the world, but certainly it is a neighborhood containing the skyscrapers of North America.

On this climb the observant naturalist will note the gradual changes of vegetation as the tree line is reached, and finally passed; noted too are the variety of animals and birds living amid the rocks and mountain plants, and the shapes of the mountains all around. But there is more to see here, or, more accurately, *not* to see. For the determined paleontologist, where every layer of sedimentary rock may hold the fossils of a treasure incarnate, this climb is a great bore. While sedimentary rocks abound, and indeed are trodden on for most of the more than six miles of this upward hike, there arc no joys of discovery that even the meekest fossil elicits. Until nearly the end of the climb the rocks on this walk are barren of life. This is not to say that life did not exist on the planet when these rocks were deposited. To the contrary, the planet teemed with life. But it was life microbial, life at the single-celled stage, life yet to evolve the intricacies of coordination in a large body. The Earth had an abundance of slime life, life dominated by bacteria. Because it is tiny and soft-bodied, without any kind of skeleton or carapace or shell, such life rarely leaves fossils. It changes the planet, but rarely leaves a fossil. It is not until the end of the hike, nearing the top of the mountain, that fossils first become abundant. But when they do, they show a sudden exuberance of shape and form, a clarion call of new complexity and unfolding diversity. It has taken a long climb to arrive at the start of the Age of Animals, just as it took our planet almost 4 billion years to achieve the summit of the animal grade of evolution.

The sun is rapidly falling when we finally attain the slopes near the top of Mount Stephens. From afar our goal looked like a small gash cut into the side of the mountain. Now, as we pant and slowly climb the steep slope of scree and talus, the gash resolves itself into a stony ledge jutting from the side of the tall pinnacle of shale and limestone that we have just climbed.

Sitting well above the tree line, the ledge is about a hundred meters long and has taken a century to form. It is composed of mounds of split and discarded shale, and sits in front of a quarry cut back into the rocky mountainside. The quarry is not a mine in the traditional sense; there is no silver, gold, or other precious minerals to be found. Instead, its treasures are small smears and impressions imprinted onto the dark shale: fossils that give the world's best clues about how the first animals on Earth lived and looked. The quarry has grown, slowly, as hunks of fissile shale have been painstakingly pried out of the mountainside by several generations of paleontologists, examined, and then discarded onto the ever-growing discard pile. This place—the quarry, the ledge, and most directly the vein of rock giving rise to it all—is known as the Burgess Shale, a place—and time—popularized most famously by Stephen Jay Gould in his book *Wonderful Life*. Earth and life scientists revere it as the most important fossil site on Earth, a site whose spectacular fossils are the world's best example of an event known to geologists as the Cambrian Explosion. The Burgess tells us that life finally reached the pinnacle of design and complexity that we call animals only after more than 3.5 billion years of evolution on our planet. It also tells us how rapidly various categories of life can appear—and then disappear.

It wasn't until very recently that the Burgess Shale fossils have been accurately dated. While some rocks lend themselves to age determination, others—especially fossil-bearing shale—are much more difficult to age. In the late 1990s scientists finally put an accurate number on the Burgess Shale—slightly more than 500 million years in age.

On a good day, when it is not snowing or hailing, and the wind is not blowing a gale, the field crews can find tens to hundreds of fossils collected from the scree and shale that has been quarried out of the mountain's side. Many scientists have now pored over these fossils and the copious ancillary data, and the result is a comprehensive view of this bygone world. We know now how it may have looked on this deep, cold, and ancient sea bottom, a place apparently writhing with life; we can discern the tapestry of this ancient world, begin to know of its carnivores and herbivores, scavengers and drifters, sessile filterers and decomposers, a menagerie all in shapes and forms so unlike the life of our world but all related, deep ancestors indeed, our roots. Nondescript and meek in this lost world wriggled small wormlike creatures that would become fish, and eventually higher vertebrates, and—so recently—us.

The fossils from the Burgess Shale are the vanguard of what has been called the Cambrian Explosion, that seminal moment in the history of life on Earth when all of the animal phyla rather suddenly appear in the fossil record. It was an event like these mountains, a huge rising from the monotony and simplicity of the microbial world that had dominated the planet for more than 3 billion years. The Burgess Shale tells us that animals did not slowly appear on the planet, but did so in an impatient rush, seizing the world in hegemony and dominion from simpler life-forms. The Age of Animals began, and is recorded here, and in so doing ended what might be called the Age of Bacteria.

It is fashionable in some scientific circles to conclude that the Age of Bacteria *never* ended, that it continues still, with we animals but an ephemeral ground cover on a planet infested and dominated by microbes. Yet this is apologia; higher plants and animals share the world with microbes, but we do so according to *our* rules, not theirs. We know from the fossil record that life first appeared on planet Earth almost as soon as it physically could have. That first life was a bacterium. Bacteria then ruled the world until the appearance of animals as witnessed at the Burgess Shale. The changeover was swift, the end of the bacterial world brutal. Our old friends the stromatolites went from dominating the ocean shallows to being eaten out of them by the new proliferation of animals. This succession was accomplished thanks to the rise of atmospheric oxygen permitting animals and higher plants on the planet for the first time. The world changed, and the low-oxygen world that had persisted for more than 3 billion years had ended.

After generations of collection and analysis, scientists have a comprehensive view of this dawn of animals. All were sea creatures, for animals had not yet crawled onto the land. Yet even at this very fossil-record beginning they included the full range of carnivores and herbivores. All of the animal phyla suddenly appear in the fossil record, a huge advance from the monotony and simplicity of the microbial world. The Burgess Shale tells us that animals did not gradually appear on the planet, but did so in an explosion of new life.

Will we leave the same way? Our triumph will end in the distant future with the end of planets and loss of oxygen, and we can begin to picture that future by imagining the Cambrian past. Let's float in a raft now above the sea bottom that will one day become hard sedimentary rocks that we call the Burgess Shale.

Many odd differences are instantly noticeable. The first, and most alarming, is that it is difficult to breathe. Second, the temperature is hot. While the air has much lower levels of oxygen than our own time, it has hundreds of times higher levels of carbon dioxide. The air and clouds look different, and even the light has a different hue. The Sun is slightly less bright, but the air is warm: the high levels of carbon dioxide have created a stronger greenhouse effect that traps the feeble rays of the Sun and ensures the warmth of the planet. The nearby shore of land is entirely brown, without trees or bushes of any kind, for land plants have not yet evolved. But there is life on the land, life in abundance; it is covered in oleaginous masses of bacterial slicks, a multihued scum that is a rancid butter of life. The land is smeared with organic material. Here and there, among this bacterial manna, a few hardy pioneers are present. A wisp of green moss. A small, tentative arthropod that will be an ancestor to today's insects, struggling across a bacterial mat. A flatworm inches through the intertidal mud. These are just the first scouts of a coming invasion, for in the nearby shallow seas vast hosts of animals and plants are taking on the evolutionary weapons, breathing apparatus, and armor that will allow them to invade the land and change it from a place of bacteria to a place of animals and plants. In doing so they will radically change landforms, river courses, soil, and even the atmosphere.

The beach is enormous in extent because the tide rises and falls fifteen vertical meters each day. The cause of this is apparent at moonrise, for the Moon is still closer to the Earth than what we are used to. Because of its closeness it seems to shoot up into the sky. The Moon's features look familiar enough because the Man in the Moon is already there, formed by huge meteorite strikes in that violent birth of our solar system already billions of years in the past. Yet even with its great size the full Moon is not brighter, and we realize that the dimmer Sun is reflecting less light from this ancient Moon. We lay back now, still fixated on the sky, and make two final observations, both dealing with the stars. Can we use the stars as a guide to our latitude on this ancient Earth? Can we fix our latitude by seeking out the North Star and the Big Bear, and hence know if we are even in the Northern Hemisphere? No. We can see the Milky Way and planets but, search as we may, no familiar constellation is apparent and, moreover, the stars are moving across the evening sky at a slightly faster pace than in our world. It can mean only one thing—the day is shorter here; the Earth revolves on its axis at a faster rate than the world we come from.

This was the beginning of the reign of animals and plants. The first trickles of animal evolution soon burst into a torrent of speciation, first filling the seas, and then spilling out onto the land, as evolution forged ever more complex and sophisticated organisms of which we are but one branch. But the Burgess Shale records more than the simple beginning. It suggests our planet's likely future. If the Earth persists long enough, there will someday be another succession of sedimentary rock strata that will, in all probability, record a mirror image of this now ancient Cambrian world: bacteria will conquer again. Complex animal and plant fossils will be overlain by a succession of strata showing ever-simpler organisms, the history of a world dying, system by system. Eventually, there will be rock completely barren of the fossils of life. If we could climb to the Burgess Shale of the future we would see fossils below but not above the quarry.

The time between the loss of higher land plants due to low levels of carbon dioxide and the loss of land animal life due to either the utter loss of oxygen or the rise of temperature will be millions of years—perhaps tens or even hundreds of millions, depending on which scientific model one believes. And in that vast amount of time the resourceful engine of evolution will be working overtime, creating new morphologies and physiology that can combat the ever-changing and more difficult physical environment. While the carbon dioxide decline will ultimately reduce the presence of plant life on the planet, its complete elimination would not in itself cause the end of life on the planet. Life will also be lost due to lack of atmospheric oxygen, and the continually increasing temperature.

TEMPERATURE AND TEMPERATURE LIMITS ARE IMPORTANT IN controlling biodiversity and the distribution of animals and plants in today's world, and it is a fair bet that they did so in the past and will do so again in the future. It is a certainty that global temperature will get ever hotter once we have passed out of the time of Ice Ages, because of the growing heat of the Sun. The future pattern of evolution and extinction will be tied to rising temperature.

The various mathematical models of the Earth's future, while differing in detail, all agree on one point: as soon as carbon dioxide levels reach the critical level that kills plants, the long, slow increase of temperature on Earth will change to a rapid temperature rise, climbing 10 to 20 degrees Centigrade in 100 million years. The length of time without plants, but with animals, might thus be as long as 200 million years, but could be as short as 100 million years. We can conjure visions of a world astoundingly different from our own, a place where fungi, algae, and bacteria form the base of the food chain, and where animals have reorganized into trophic guilds and communities completely foreign to us. This is a world as yet never envisioned by any form of popular culture, and yet it will be a reality on Earth for many millions of years. Some things are indeed more fantastic than science fiction.

Let's travel then from the bizarre seas of the Cambrian Explosion some 530 million years ago to a time equally distant in the future, and watch the progression of events that might take place following the loss of land and sea plants.

If asphyxiation immediately kills off animals following the loss of plant life because oxygen disappears and lacks replenishment, the following scenarios will not take place. (We'll note that nervous tissue, needing the most oxygen, will be the first sacrifice to the new world order: brains and nervous systems will become obsolete, and the world will devolve into ever greater stupidity, with less complex sensory organs and behavior. A cheerful thought: brave and stupid new world.)

But if animals and oxygen linger, rising temperature will play a key role. As mean global temperatures exceed 38 degrees C (100 degrees F) (the temperature of a hot tub) the planet will begin to die at the equator, and multicellular life will have to migrate toward the poles. (Mean global temperature is not the hottest temperature, but the average temperature of the whole planet, evened out for days and nights, and thus a mean temperature of 38 degrees Centigrade will result in considerably higher noon temperatures at the equator.) The continents in the vast middle parts of the world will become largely devoid of surface life due to the high heat. There will be no snow and no ice caps.

The oceans, while less affected than the land, will nevertheless be permanently changed. The reduction in nutrients because of the loss of plants on land will reduce oceanic productivity, and thus in all probability disrupt almost all oceanic community food chains at all water depths, even those on the deepest ocean bottoms. Only those communities clustering around deepwater volcanic vents with chemical energy sources will be immune. Coral reefs will forever disappear from the planet, unless some new form of low-light internal algae of the kind that allows corals to grow large skeletons evolves for deep, cooler water.

The seas themselves will in all probability change in color, since there will be a reduction (and in some places an elimination) of green plankton, as well as an enormous increase in the amount of sediment found within the upper levels of the water. With the loss of plants on land, vast amounts of topsoil will erode into the sea. As vast deserts reclaim the land surface, much as they did in the long-ago Precambrian era, great dust storms will carry more sediment into the oceans. The sea will become browner.

Yet it will not be a desert world of arid dryness—far from it, in fact. The high temperatures will ensure that the air is charged with moisture. Imagine the Amazon region on the hottest, stickiest day of its year, but without plants to lend shade. That wet, prickly heat will be a fact of the Earth.

When global temperatures reach the 40 degrees C (104 degrees F) range (hotter than the highest temperatures that much of the Earth now experiences), dramatic changes should occur to animals. First of all, this will be a time of phenomenal evolutionary change. The rate of heat increase will be slow enough, with perhaps a 1-degree rise every ten thousand to one hundred thousand years, to allow evolutionary adaptation. Among land animals such changes will occur in their behavior (such as spending their foraging time at night), their internal organs, and their body plans. Secondly, it will be a time of extinction, as those organisms that are not capable of rapid evolutionary change disappear. The zoo of life will be a rapidly flickering movie of new animal species desperately adapting new strategies to keep up with the rising heat—or going extinct. Biological communities will no longer be long-lived or stable. There will be ecological chaos amid a dizzying evolutionary procession of forms scurrying about the *Titanic*'s deck, all trying—to no avail—to evolve life jackets or, in this case, a foolproof method of biological refrigeration. Even burrowing ever more deeply into the soil or the bottom of the sea will eventually be to no avail, as the heat cooks the Earth in its humid oven.

Let us imagine a scene from this world. Because whipping, hot wind lashes the land surface and makes surface life virtually untenable, armies of animals have evolved an underground lifestyle. The food chains no longer are based on the higher land plants now long extinct, but on bacteria, lichens, and fungi. The Earth is now a haven for many organisms once restricted to hot springs. There is still surface water on land, for the high heat has created a humid atmosphere and still-abundant rainfall, allowing the presence of ponds and even lakes. Yet these are fetid bodies of water in which rainbow hues of bacterial slicks abound. The microbes are quite content in the new, higher heat, and it is their biomass that now sustains the rest of the terrestrial biota.

We know from various studies that animal life in the present day has as its maximum temperature about 45 degrees C (113 degrees F): at this temperature the cell's mitochondria cease to function. (These are the sausage-shaped organelles in a cell that convert food molecules to a form the cell can use.) In our future world most animal life that still exists will be clustered at the poles, for a mean global temperature of 45 degrees C means that the equator will be far hotter still. Even near the poles, surviving animals will be nocturnal, hiding from the murderous Sun. We might expect animals to hibernate during the long summers of constant daylight, waiting for the winter months of constant darkness. We can imagine evolutionary experiments of molelike creatures, lurking in tunnels underground. As the temperatures continue to rise, underground fauna might become the only fauna, with all of the land surface now sterilized of the more evolved cells called eukaryotes, leaving behind only primitive bacteria. Gradually the world (at least on land) will return to its Precambrian state.

As global temperatures reach 50 degrees C, wholesale extinction will be taking place on land. In the oceans the situation will be more complex and harder to model. In such high heat, ocean seawater will certainly warm to levels not experienced since the Hadean era of more than 3 billion years ago. But life in the sea will be more difficult to kill off. The heating and insulating properties of water are very different from those of air, and there is already a well-known range of marine organisms capable of dealing with these temperatures. Many more will evolve in the time when the pulse of change on Earth quickens with the rising heat.

As temperatures continue to rise above the 60 degrees C (140 degrees F) range, the land world will begin to resemble what it once was a billion years ago, before the first animals and plants had evolved. There will still be a remnant of life on land, and somewhat more in the sea. But not much. Bacteria, algae, and fungi will be the kings of creation now, along with lesser armies of protozoans and perhaps lower invertebrates such as nematodes and flatworms. Diversity will be found among sand grains, in the rainbow bacterial slicks, and in the clinging lichens and mushrooms, persisting in the high humidity. Perhaps the stalked mushrooms will be the only trees of this world, the half-inch-tall redwoods of a once forested land.

The time will come when global temperature reaches 70 degrees C (160 degrees F). Except for bacteria, life will have been extinguished from land. Only the seas will still harbor some reserves of higher life. Until, that is, the oceans themselves begin to disappear.

# · 8 ·

# THE LOSS OF
# THE OCEANS

WE HAVE REACHED A TIME, PERHAPS 1 BILLION YEARS FROM now, when the mean global temperature is near 70 degrees C (160 degrees F) under a Sun that is 10 percent brighter than its present value. In such an environment we humans could not exist for long without refrigeration of some sort. The heat would kill us, just as it will have killed all higher animals on the land and most in the sea. Yet even this bleak planet has some familiarity to us, compared to the Earth that is to come. For now the oceans will begin to evaporate away and be lost to the cold vacuum of outer space. With this stage the planetary systems for life—as we know them—will be either in their final throes of existence or already stopped. The carbon cycle will be so transformed as to be nonexistent. The circulation patterns of other nutrients will be winding down. The planetary thermostat system will have closed down. As an abode for life the planet is dying. Not like death in the movies—a shot, a few dramatic last words, and down go the eyelids in a grandiose curtain close—but as most organisms die, as Ruth Ward died, slowly, with systems shutting down for a last time.

It is a familiar fantasy that the sea vanishes, baring its secrets and treasures. Sunken ships, writhing fish beyond number, and newly exposed monsters are suddenly exposed to the glare of Sun and human scrutiny. What we don't realize is that in the distant future this will indeed happen, and its reality will be stranger than our fantasy. Huge new areas of nutrient-rich sea bottom will become land, but there will be no plants to colonize this rich source of nutrient, only bacteria, which will coat all with oozing slicks and layered stromatolites. As the seas continue to recede, first dropping hundreds, and then thousands of meters into the deepest parts of the ocean basins, the entire geography of the world will change into one where land dominates the world, where oceans have become seas, and where mineral salts will mark where waves once lapped.

When perhaps only a quarter of the ocean volume is left, an entirely new landform will emerge. First breaking the surface as smoking and belching volcanoes, entire ranges of volcanically active mountains will divide the remaining seas. These great mountains, only tens of kilometers wide but thousands of kilometers long, will be the tops of the midocean ridges, the places where continental drift creates new earth material along spreading centers. Today they emerge in rare places, such as Iceland. But the dropping oceans will expose all that are still active in that far distant time.

The loss of so much water into the heated air will drastically affect the composition of the atmosphere. High in the atmosphere water molecules will be stripped apart into their component atoms, with light hydrogen floating off into the darkness of space and heavier oxygen left behind. The oxygen-starved world that doomed all animals will now reverse itself and become oxygen-rich, but only during an era of life-destroying heat. Eventually the weight of oxygen on the land will be crushing, producing Venus-like pressures, at least for a while.

Terrible storms will lash the planet, and the constant lightning strikes amid this ever-richer oxygen atmosphere would seemingly be a potential recipe for explosive fires. But in this far future world the forest will already have vanished. There will be nothing to burn, except, perhaps, for nitrogen in the atmosphere, which will be transformed into nitric acid, producing planetary acid rain.

The dwindling seas will become ever more salty, and the nitric acid rain will add to it to produce a highly poisonous brine. Even hardy bacteria that thrive in hot water will be hard-pressed to survive the poisonous stew. At the site of the Earth's hydrothermal systems, where seawater is taken down into the magma near the ocean-spreading centers, this salty and acid fluid will cause whole new suites of metals and minerals to form at the hydrothermal hot spring sites.

When this process is complete, our planet—our beautiful, water-covered planet—will be like some lovely painting suddenly stripped down to bare canvas, its ocean basins vast plains of salt except for lingering pools and ponds. Our treasured planet—blue for billions of years—will turn a pinkish-brown. California's Death Valley provides a hint of what this awful future will be like. Bad Water, a small saline lake in one of the hottest places on Earth, is hot, saline, choked with sediment, and home only to bacteria. This is what the last of our planet's water will be like, and even Bad Water is far purer and cooler than the noxious stew the oceans will leave behind. Yet how can the mighty oceans, which average nearly two miles deep and cover nearly 70 percent of our planet, simply disappear?

WHILE THE OCEANS APPEAR TO BE CONSTANT, THEY ARE, IN fact, already being lost to space, but at a rate far less than in the hell of the future. They are evaporating to the cosmos at the rate of a millimeter every million years, a pace so slow that it would take trillions of years to make a serious dent in the planet's water supply. But that rate will radically increase as global temperature increases, so that within 1 billion years from now, the Earth's entire oceans

will begin to seriously stream off into space, a victim of the inexorable increase in the brightness of the Sun. The process begins in a
billion years with what is known as the *moist greenhouse* effect. If
this is not totally successful in annihilating the oceans, the much
more severe *runaway greenhouse* effect will kick in 2.5 billion years
later when the Sun's brightness has increased by 40 percent. It is
generally believed that the moist effect will succeed in driving the
oceans into space all by itself. Curiously enough this act may save
the world for possible long habitation by heat-and-salt-loving
microbes. For billions of years the oceans have been one of Earth's
most important assets but as the Sun becomes increasingly bright
they become a lethal liability. For any life to survive well into the
second half of Earth history the oceans must go! Microbes will
surely survive ocean loss by the moist process and they might survive for billions of years afterward in an ocean-free world. If the
moist process is not quick enough, however, the surviving oceans
and the vapor that they inject into the atmosphere will produce
thermal runaway—hell on Earth—and conditions that no life could
imaginably survive. Although animals will not survive to see the
outcome, the fate of the microbial world will hinge on Earth rid-
ding itself of nearly all of its surface water.

The fate of the oceans and the timescale of their destruction
have only recently been investigated, thanks to the work of James
Kasting at Penn State University. Yet the idea that Earth's oceans
are in risk of disappearing is at least a century old. Some of the earliest quests for life on Mars are linked to ideas of water loss.

In the beginning of the twentieth century, astronomer Percival
Lowell built an observatory on Mars Hill in Flagstaff, Arizona,
renowned for its dark and stable skies. He then began a study of the
surface features of Mars. His main telescope, a twenty-four-inch-
diameter Alvin Clarke refractor, had the highest-quality optics of
its time. Even at Mars Hill, however, the sky is not perfect, and the
fine details of the Martian surface would wobble and throb in and
out of focus. For brief periods of time the details of the planet

would be sharp, and then they would degrade. When conditions were optimum, fine subtle details could be seen on the surface of Mars. The best technique of the time was to look through the eyepiece, record an instant of focused observation in memory, and then sketch its surface details on a piece of paper.

Hour after hour and night after night, the observations continued. Subtle patterns of dark and light, nearly a hundred million kilometers away, were viewed through our atmosphere of constantly moving air. The sketches were fabulous, the best that had ever been made of the Planet of War. The best images were seen when Earth and Mars were closest, at opposition, a planetary alignment that occurs every eighteen months. Because Mars is tilted on its axis like Earth, seasonal variations could be seen. The most conspicuous were the seasonal growth and shrinking of the white polar ice caps.

The ultimate product was a globe that mapped the intricate details of Mars. The most prominent features were dark lines that formed a weblike network covering the Martian surface. Lowell called the lines canals, and his suggestion that they might be made by Martians created a worldwide sensation. He reckoned that the canals were dug to provide irrigation for the dying planet. Mars is a rust-colored, desiccated planet and does not have visible bodies of water. The astronomer reasoned the lines were dug as desperate acts of a dying civilization on a world that was losing its water to space. The canals would transport water from wet high latitudes to dry regions at moderate and equatorial latitudes. During the Martian seasons there was a "wave of darkening" seen in the telescopes that seemed to spread from the poles. It was theorized that this was due to the spring growth of crops irrigated by seasonably available water.

Lowell saw the dying Mars as a prelude to the future of Earth. He recognized that the life of a geologically active planet is dependent on its mass. Lowell thought that the Moon, as a small body, had already evolved, leaving a dead planet covered with dried seabeds. Mars, intermediate in size between the Earth and the Moon, was in its final days as an abode of life. Earth would be next to go. It was well known at the time that marine fossils were found at high altitudes on Earth, and that there was abundant evidence of much higher sea levels in the past. Sedimentary rocks and marine fossils are found from the western flanks of the Appalachian Mountains to the front range of the Rockies. There was clear evidence that seawater formerly filled the interior region of the United States. To Lowell—working in an era well before modern understanding of plate tectonics and geology—this was evidence that ocean levels were dropping and that it was just a matter of time until the Earth became a desert planet void of oceans. In the early 1900s the age of the Earth was still unknown, and it was believed this process might occur on a relatively short timescale. Observed ancient shorelines above sea level indicated that the Earth's oceans had dropped even during human history. If Earth was rapidly losing its oceans to space, Lowell concluded that Mars would have lost its oceans much faster because of its smaller size and proportionately lower surface gravity.

As history would tell, Lowell's canals were not real. Yet although Lowell was wrong in the case of Mars, the imagined effect is real. The study of Mars in a real sense may be a "back to the future" lesson for Earthlings. Oceans have a limited lifetime on any Earth-like planet, and in time they are inevitably lost to space. If intelligent life survives long enough, Earthlings of the distant future will surely consider Lowell to be a prophet.

THE LOSS OF EARTH'S OCEANS CAN ACTUALLY BE OBSERVED IN real time. This was first seen by a remarkable telescope placed on the surface of the Moon by the crew of *Apollo 16*. The telescope was about the size of a picnic cooler and looked like a very fancy metal box resting on top of a camera tripod. It gathered its images

An ultraviolet image of Earth taken from the surface of the Moon by the *Apollo 16* astronauts. The glow on the left, the direction of sunlight, is Earth's geocorona caused by the fluorescence of hydrogen atoms escaping the planet. The image is a vivid illustration of the slow loss of our oceans to space.

from the ultraviolet spectrum of light, and it could detect the fluorescent glow of Lyman alpha radiation of hydrogen gas. The glow is caused by electrons falling back down to their lowest energy levels in excited atoms of hydrogen gas. Because ultraviolet light is invisible to our eyes, it cannot be seen from Earth. Even special telescopes can't detect it because the atmosphere screens this kind of radiation. The Lyman alpha glow can be observed from space only with special instruments. The Moon provided an ideal vantage point for viewing the whole Earth, and *Apollo 16*'s "Carruther's camera" provided stunning images, which show Earth to be surrounded by a ghostly halo of fluorescing hydrogen gas. It was as if the Earth was emanating a magic aura into space. The glow was not uniform but extended most fully toward the direction of the Sun, the energy source that excites the glow.

Optical images of Earth show a beautiful blue planet with only a paper-thin layer of hazy atmosphere. Its wispy glow can be seen to extend only a few tens of kilometers before it is lost in the blackness of space. In the ultraviolet range, however, the Earth's strange Lyman alpha aura extends tens of thousands of kilometers outward, a sphere of escaping gas much larger than the Earth itself. The gas density is ultra low, even lower than can be obtained by the best vacuum pumps on Earth, but it is enough to emit the powerful ultraviolet glow. Comets are also surrounded by large Lyman alpha halos for exactly the same reasons—like Earth they are losing hydrogen to space. Most of the glowing hydrogen seen around the Earth is from the oceans, and it is dramatic and sobering evidence that the planet's water supply is headed for a catastrophe. The loss is slow, but the ultimate effects are remarkable. The oceans will almost totally evaporate, leaving the ocean floor a huge wasteland encrusted with salt.

The loss of the oceans starts when water molecules leave the ocean through evaporation or the bursting of a bubble in a wind-blown wave crest. Most linger in the lowest 12 kilometers of the atmosphere—the troposphere that contains our weather and clouds—and return to the surface as precipitation. But some are carried upward by convecting air to the top of the atmosphere and are split into hydrogen and oxygen. The light hydrogen ultimately escapes into space while the much heavier oxygen stays in the atmosphere.

Why is ocean water escaping today at only a rate of a millimeter per million years? Because the frigid stratosphere is too cold to hold and transport much water. As water vapor ascends, the atmospheric temperature drops at a rate close to 9.8 degrees Centigrade per kilometer of altitude. Where the troposphere ends and the stratosphere begins, the temperature has dropped to about −65 degrees C, depending on the latitude and season. This helps conserve our oceans because the amount of water vapor that can be retained in air depends on temperature. On a hot August day in Houston with 100 percent humidity, the air is more than 6 percent water vapor. When it is 40 degrees below in International Falls, Minnesota, even on a damp day the frigid air can only be one ten-thousandths water. Similarly, at the top of the troposphere, only ten molecules in a million are water vapor. This frigid dry air is a cold trap that prevents water from rapidly ascending to the highest altitudes where it can be lost to space. As a result, the present loss rate is only a meter of ocean in a billion years.

Consider the travels of a single molecule of water. Once it climbs into the dry, frigid stratosphere, it is in a region of the atmosphere where the temperature is fairly constant and even rises somewhat with altitude. It is also free of turbulence, a fact well known and highly appreciated by air travelers. In the stratosphere, our air molecule is carried upward not by local updrafts but by global-scale air movements that occur between polar and equatorial regions.

If the water molecule is carried high enough—above 20 kilometers altitude—the molecule is above the ozone layer that shields the lower atmosphere and creatures of Earth from the effects of ultraviolet sunlight. In this region the molecule can be broken into oxygen and hydrogen by ultraviolet radiation, but then it quickly recombines to form water again. As upward migration continues, however, the ultraviolet light becomes more intense, and the collision rate that leads to water reformation becomes less and less. At altitudes above 150 kilometers the molecule is permanently split by ultraviolet light and the hydrogen and oxygen atoms travel at different speeds due to their different masses. The heavy oxygen is too slow to acquire the escape velocity of 11.2 kilometers per second to leave the gravity of Earth, but a tiny fraction of the light hydrogen atoms manage to fly into space.

Most hydrogen atoms do not reach this speed, and as such they travel upward on a high arching trajectory before being struck by other atoms and molecules and eventually falling back to our planet. Still, over billions of years, more and more "slip the surly bonds of Earth." It is just a matter of time when, by chance, the hydrogen has a great enough speed and is traveling in the right direction at a high enough altitude. In some cases the hydrogen picks up electrical charge and gets an additional electromagnetic boost from Earth. As our fugitive atom is lost to space, it produces the Lyman alpha glow that was observed from *Apollo 16*. Once beyond the region of this glow, it will be carried by the solar wind streaming out of the Sun, passing the orbit of Pluto in less than a year and then traveling on into interstellar space. Ultimately the atom will be captured by gravitation in other stellar systems and be recycled into future stars and planets. Atoms are the one part of the Universe that live forever, and are eternally cycled in out of organisms, oceans, atmospheres, rocks, stars, planets, and the "emptiness" of deep space.

THE EARTH'S LOSS OF WATER MAY BE SLOW TODAY, BUT IT WILL dramatically increase in the future because of the increasing brightness of the Sun. As the Sun gets brighter, the Earth's surface temperature will climb and the height of the troposphere (or the lower atmosphere where our weather occurs) will ultimately rise from 12 kilometers to over 100 kilometers. When global surface temperatures reach the disastrous 60 degrees C that will end most animal life, the water vapor concentration in the troposphere will be 20 percent and the planet's "cold trap" will essentially cease effectively to keep the oceans from streaming into space. This will be the start of Kasting's moist greenhouse mentioned previously. The main feature limiting the rate of water loss at the top of the atmosphere today is the slight concentration of water vapor in the stratosphere, which in turn is determined by the water vapor content of the top of the troposphere. As this concentration climbs as the atmosphere heats up, the troposphere will swell and the loss of oceans will rapidly accelerate.

This changeover will spell not only the loss of oceans, but the creation of a newly dramatic and terrifying sky. The hot humid air rising up from the roasting Earth will become unstable to altitudes of 100 kilometers and lead to storms more violent than the world has ever seen, a roiling atmosphere that has no parallel in human experience. Thunderheads will rise far higher above the Earth than they now do. Huge hailstorms will pelt the upper atmosphere regions as ice forms from the great quantities of water rising upward, yet none of this giant hail will ever hit the Earth. All will melt on the way down. Earth's oceans might be lost in only a hundred million years, much like the draining of a bathtub although there is a chance the process could be slower. On the timescale of a human life, the water level would drop only a few millimeters, an unperceived amount. But compared to the long history of our planet, the seemingly eternal ocean will evaporate relatively quickly, constantly decreasing the fraction of Earth covered by seas and increasing the amount of land. This alone will accelerate the process, for the enlarging land area that absorbs more sunlight will combine with the greenhouse effect of increasing water vapor in the atmosphere to, in a vicious circle, cause ever more heating and temperature rise.

As the oceans drop first by hundreds and then by thousands of meters, they will become increasingly hot and saline because even though the water leaves, the salt will be left behind. This radical change in salinity will have geological as well as biological consequences. Huge areas formerly covered by the sea will now become regions covered by thick sedimentary deposits of salt minerals, creating immense beds of halite, barite, and gypsum. Eventually these great salt flats will become nearly continental in size in some areas. For surviving marine animals, it will be a sea where the salt requires ever more evolutionary adaptations to deal with this new poison. Better kidneys and a better excretory system will be required.

What will live in the hot salty oceans? It depends on when you look. As the oceans recede, there will be a protracted series of extinctions of marine animals and mammals. The oceans were warmer during the Mesozoic and Paleozoic eras in the past when a higher abundance of carbon dioxide in the atmosphere led to great greenhouse warming. When the conditions cooled from these warmer, essentially reptile-friendly times, a number of new species evolved in the cooling oceans. As the oceans warm again, perhaps history will reverse itself and species that most recently evolved will be the first to be lost. This will include krill, whales, porpoises, and then sardines, tuna, and salmon. Lobsters, jellyfish, anemones, and coral will follow the unstoppable countdown to doom. As corporation Earth executes its final downsizing, the loss of each species will provide new opportunities for competitors and doom for predators. As has always occurred in the past, the ecology of the living Earth will adapt, but as the sea becomes ever more hostile, the complexity and diversity of life will continue to decline on a slippery and ever downward path. Eventually, life in the oceans will revert to its microbial beginnings, and, once again, our past will become the future. Perhaps this fate will be delayed while new, formidable superspecies adapt to the hot and salty seas, but as the oceans continue to drop, eventually they will all die off. In the end the ocean will be dominated only by bacteria that can thrive in its hot, salty, unpalatable brew.

In the final stages the oceans may look like wildly colored ponds scattered across vast plains of salt, resembling those used for extracting salt from seawater in the south end of San Francisco Bay. As these ponds dry in the brilliant California sunshine, they become a kaleidoscope of vivid colors ranging from bright red to pastel green. This is because of the microorganisms thriving in incredibly salty water. Eventually, of course, they become vast ponds of salt. So too will the oceans. The last life they contain will be halophiles, or microorganisms that live in tiny water inclusions in salt.

While the ocean floor has mountains and trenches, for the most part the vast abyssal plains are flat. When they dry they will become deserts of salt with numerous scattered ponds of brine. The ocean is 3 percent salt and so the drying of the 4,000 meters of ocean depth will leave more than a hundred meters of salt, washed by rain from high to low areas. The final state of the Earth's formerly great oceans will be desolate and nearly endless salt flats, similar to the floor of Death Valley in California.

Long after the ocean is gone, scattered ponds of salt-saturated water will remain, lined with crystals of salt. Water will continue to be lost from the top of the atmosphere, but there will always be some outgassing of new water from the Earth's interior from volcanic activity. Until the Earth's surface is heated to extreme temperatures there will always be someplace on the surface of the planet where some precious fluid can remain and nurture the planet's last surviving organisms. But when the temperature finally reaches 374 degrees C, there will be no water. This is the critical point above which water cannot remain a liquid at any pressure. When this point is reached, the planet will become waterless and the last life based on water will certainly perish.

When will this superhot world happen? If the moist greenhouse effect operates fast enough, life will survive into the postocean world. An ocean-free world will be a drier one with less water vapor operating as a greenhouse gas in the atmosphere, and thus cooler, allowing remnant bacteria to survive in a dry but not entirely waterless world. With no greenhouse effect at all, the warming Sun would increase surface temperatures only about 30 degrees C over the next 6 billion years.

But if the evaporation process of the oceans is too slow, Earth

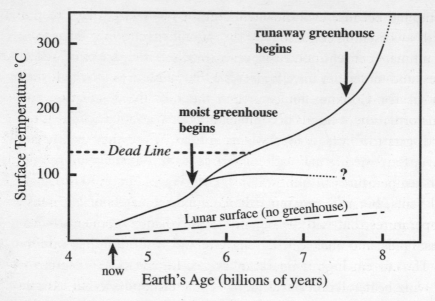

The loss of Earth's oceans will determine the long-term fate of even simple life on Earth. Our crude estimates of the possible future temperature history of Earth are made combining both Jim Kasting's models and the future brightness of the Sun. If the oceans are not lost to space, atmospheric vapor will drive Earth to a runaway greenhouse and the severe temperature path leading upward. If the oceans are lost before the runaway begins, Earth's surface might follow the more moderate lower curve—for a while. The evolution of the lunar surface temperature with time is shown to illustrate the critical role that greenhouse effects have in severely amplifying the effects of solar heating.

will become too hot for life while the oceans are still here. If the oceans survive too long, then rising temperatures will drive Earth into a runaway greenhouse effect so calamitous that atmospheric temperature could rise to over 1,000 degrees C and the planet's surface will actually melt. Kasting has estimated that this could occur in about 3.5 billion years when the Sun becomes 40 percent hotter than it is now.

With the disappearance of all water, the last organism will finally be gone and with it all evolution and ecology. When will the end come? We can predict the sequence of future events better than its timing, because evolution will continue to try to adapt to our Earth's worsening conditions. If the date of execution is uncertain, the ultimate cause is not: rising temperature. Life is based on molecules, and molecules have limits to the temperatures at which they can survive. Cooking dinner is fundamentally the destruction and transformation of bonds in complex organic compounds. When the Earth gets too hot, its life systems will cook, degrade, and break down. Extreme life may be remarkable in its ability to survive the high temperatures of Yellowstone hot springs or deep hydrothermal vents, but it is not remarkable enough to endure the hellish temperatures that will come. Life's most basic components—proteins, lipids, and nucleic acids—have relatively modest thermal limits. The current high temperature record for life is 235 degrees F. Evolving heat-tolerant bacteria, like the thermophiles that exist in today's hot springs, will make life's last stand and may advance this record by tens of degrees or more, but surely not by hundreds of degrees. Probably the last organism will be an autotroph, an organism that synthesizes organic compounds from inorganic sources and lives below the earth's surface. Its last stand would likely take place beneath the surface of a tall mountain in the polar regions, where temperatures would stay the coolest. When its environment reaches 300 degrees F, it too will perish.

If the oceans evaporate relatively quickly and a runaway greenhouse effect can be avoided, perhaps life will persist until almost the end of the Sun's life as a main-sequence star. Life, in primitive fashion, could survive another 6 billion years.

Alternately, Earth will become a copy of Venus, where the surface temperature of 450 degrees C is hot enough to melt tin and liquid water cannot exist at any pressure. We have no information about the early history of Venus, but presumably it had an ocean of water that it entirely lost to space because it was closer to the Sun. Even if Earth does not experience Venus's runaway greenhouse effect from the loss of its oceans, it will experience one very late in its history as the Sun continues to warm. When solar heating

becomes too intense, the vast amount of carbon locked up in limestone and other carbonates will eventually be released into the
atmosphere, producing severe carbon dioxide greenhouse warming. No organism will witness this, however. When rocks are hot
enough to decompose limestone, billions of years from now, the
Earth will already be a barren planet circling a red giant star.

THE LOSS OF THE OCEANS WILL BE THE SINGLE MOST DRAMATIC
step in the annihilation of life on planet Earth. We essentially carry
seawater in our veins, as blood, and although long removed we
humans, like all animals, are reminded of our marine ancestry
through myriad biological ways. There is perhaps no vision more
apocalyptic: the blue-green planet Earth turned brown, its surface
water gone, most of its life extinguished.

As we discussed in an earlier chapter, the importance of plate
tectonics has long been recognized by geologists as central to much
that we observe and experience on the surface of the Earth; it is the
central paradigm of geology, explaining in a single overarching theory properties of the Earth ranging from the positions of continents
to the existence and placement of earthquakes. But increasingly the
theory and phenomenon of plate tectonics has become a central
area of interest to astrobiologists, for just as Earth is the only planet
in the solar system with advanced life (and probably any life), so
too is it the only planet with plate tectonics. This may not be a coin-
cidence. For our purposes we can also propose that plate tectonics
might end relatively soon in Earth's history, and when it does, there
is a fair likelihood that all life on Earth will end soon after. The loss
of the oceans might be why plate tectonics ends on planet Earth.
And with the cessation of plate tectonics, the last existing life on the
planet is doomed once again.

Two different processes could bring about the cessation of plate
tectonics. The first is the loss of the oceans. The second and quite

independent cause will be the decline of heat from the interior of the Earth until there is no longer sufficient energy to drive the convection cells that motor plate tectonics.

The first of these processes, a loss of oceans, would stop plate tectonics for a most subtle reason: it appears that a change in rock composition brought about by adding water to new lava enables rocks to dive down into the Earth at its subduction zones. As we explained earlier, these are the long linear regions where ocean crust is driven deep into the Earth. It descends not so much by being pushed down, but by sinking down through gravity. As the basalt created in spreading centers moves away from its birthplace it is light enough initially to float on the mantle, but over time it gains heavy freeloaders—piles of dense igneous rock known as gabbro, which attaches to the base of the basalt. The basalt now just barely floats, and as it cools it gets heavier. Given any good excuse it sinks back into the Earth, descending as deep as 650 kilometers— but only if it can bend enough at plate boundaries to bend downward. This requires water.

As we've explained before, the ocean plate moves away from its birthplace at the spreading centers, and it gradually changes composition: it has water added to the crystal structures of key minerals. In other words, the basalt becomes hydrated. Without water, the lithosphere (which is the plate of plate tectonics, the rigid surface region composed of the crust and uppermost part of the mantle) is too strong and brittle to be subducted. With water, the ocean floor becomes flexible enough for plate tectonics to operate.

Even in the absence of water, plumes of heated hot magma may rise to a planet's surface, but they will simply thicken into a stiff, stable surface. Venus and Mars both lack subduction zones, and thus plate tectonics. Although both might have the internal mantle convection necessary to move surface plates, the surface itself is composed of "strong" rock that cannot move. The crust on these planets, because of its thickness and strength, is now immobile. The lack of water on both of these plates may be the reason why this is so. Since both may have had liquid water in their past, and crustal composition similar enough to that of Earth (where plate tectonics does take place), we may find that both Venus and Mars once *did* have plate tectonics—and perhaps lost it when both planets lost their liquid water. Yet heat continues to flow upward, and in the case of Venus, this caused the entire surface of the planet to melt about a billion years ago. On Earth, the viscosity of our crust allows mountain formation, nutrient and carbon cycling—and life.

Like the stopping of a human heart in a long-lived individual, the stoppage of plate tectonics will have everlasting effects. Linear mountain chains will stop forming, and the ocean basins will fill with sediment eroded from the continents. The planet will become flatter. Like the crust of Venus and Mars, the crust of the Earth will thicken, and as it does so, heat will build up. It is quite likely that what happened to Venus will happen to Earth: a melting of the crust on occasion. Again, this would doom any surviving life, of course, and is one more way that loss of the oceans will almost certainly cause the final extinction.

An end to plate tectonics because of the loss of oceans could happen between 750 million and 1.2 billion years from now. But Kevin Zahnle of NASA Ames in California has raised another possibility of an earlier, catastrophic end to Earth's geologic processes: a cessation of plate tectonics brought about not from the absence of water, but by a slowing of the Earth's interior heat flow. This could end life more quickly than might otherwise occur.

It is not solar but geothermal energy that runs plate tectonics, and, unlike the Sun—which is getting warmer through time—the energy running the plate tectonic engine is running out of fuel. The radioactive decay in the Earth that generates heat declines over time. Zahnle suspects this could stop the movement of the plates in as little as 500 million years, long before the anticipated loss of the oceans. What would happen if plate tectonics ended before the oceans were gone?

If Earth's tectonic plates did suddenly stop their movement, sub-duction would no longer occur at the contacts between colliding plates. Mountains—and mountain chains—would cease to rise. Erosion would begin to eat away at their height. Eventually, the world's mountains would be reduced to sea level. How long would it take? The problem is a bit more complicated than simply measuring average erosion rates and calculating the number of years for the mountains to disappear, because of the principle of isostacy. Mountains (and continents) are a bit like icebergs—if you cut off the top, the lighter bottoms rise up relative to sea level. So too with mountains. The continents would rebound upward as they lost the weight of their peaks. Eventually, however, even this isostatic rebound effect will be overcome by the extent of the erosion.

Meanwhile, the eroded mountains would be carried by river and wind into the oceans, displacing seawater and causing the level of the sea to rise. Calculations by various geomorphologists—or scientists who study the shape of the land—suggest that there is a possibility that the entire Earth would once more become covered by a global ocean. It would be much shallower than the oceans of today, of course, but global in extent nevertheless. Our planet would have returned to the state it showed 4 billion years ago, so early in its history: a globe covered completely (or nearly so) by ocean. And with the continents awash, the Earth would witness a mass extinction more catastrophic than any in the past. All land life would die off under the lapping waves. Paradoxically, the increase of ocean area would probably also be accompanied by extinctions in the sea. Ocean life depends on nutrients, and most nutrients come from the land as runoff from rivers and streams. With the disappearance of land, the total amount of nutrients (although initially higher as so much new sediment enters the ocean system) would eventually lessen, and with fewer nutrients, there would be fewer marine animals and plants. How long before such a waterworld would be achieved? Tens of millions of years would be required for the mountains and continents to erode to sea level, but mass extinction would ensue long before. And if this occurred in a time of high temperature, it would only accelerate the loss of the oceans to space. Again, a salt-covered planet would be the result.

In either case the time of plate tectonics—the "heartbeat" and circulation of planet Earth—is limited. Its end will mark a radical change of the Earth from the planet we know to one that we do not. Just as the breakdown of one of Ruth Ward's organ systems contributed to the failure of them all, so are Earth's systems interconnected. Rising temperature, evaporating seas, and slowing plate tectonics will all combine to turn our planet into a new hell, which will finally extinguish the long history of life.

· **9** ·

# RED GIANT

the conditions of its earliest history. We are billions of years in the future. The plants have withered away, the animals have died off amid rising temperatures, the oceans have been lost to space, and plate tectonics has ended. A huge thick atmosphere has raised global temperatures to above 370 degrees C. There is no more water on the planet. All life, even microbes, has been extinguished. The evolutionary struggle to survive is over. The planet is much like it was soon after its birth—hot, inimical, sterile.

The pink, lifeless, rock-covered planet with a crushing, high-pressure oxygen atmosphere still swings around the Sun. Where once there were oceans there are but huge deserts of rock salt. The long linear mountain chains are gone and the only peaks remaining are huge, solitary volcanoes spewing lava and brimstone into a thick atmosphere of poison. Earth has become the worst of Mars and Venus, a planet sterilized. All that remain are fossils. There is no hope of any further life on the planet. But it is not the end of the world. Yet. The next events will not come from the Earth. They will come from the Sun. At this point, after billions of years of a relatively stable existence as a main-sequence star, the Sun will evolve through a dramatic series of phases as it enters the final stages of its life as a bright star. It will change from its familiar yellow color and current size and shape to become a red giant and threaten the very existence of our planet.

Yet before that happens, another drama will play out. Our home galaxy, the Milky Way, will collide with another galaxy. While this might not directly endanger our solar system, when the collision is over it is likely that our galaxy will be transformed into something quite unlike it is at present.

We are streaking at this moment toward a likely impact with a massive spiral galaxy very similar to our own. On a dark night far from the glare of city lights, you can easily see it on its free-fall approach toward us: a fuzzy patch about the size of the Moon lurking in the constellation Andromeda, halfway between Cassiopeia and the southeast corner of the great square of Pegasus. Although not at all spectacular, its ghostly glow clearly stands out in a black sky filled with stars.

This is M31, named by the French astronomer Charles Messier in 1771. Messier was eagerly searching for comets and made a list of useless fuzz patches in the sky that did not seem to move and were therefore not comets. M31 was the thirty-first fuzzy object that Messier cataloged. The astronomer's annoying object is much more than just a patch of light in the sky, however. With a good pair of binoculars and very dark sky, it can be seen to be a neighboring galaxy nearly 5 degrees across. In a little over 2 billion years from now it will be 20 degrees across, or about the size of a volleyball held at arm's length. In a large telescope, M31 is a truly grand sight of spiral arms, gas, and dust. You can even see smaller galaxies orbiting around it. Due to its spectacular spiral structure and closeness, it is a classic galaxy whose image is included in almost all textbooks on astronomy. Because we are embedded in our own Milky Way and cannot directly see our own galaxy's structure,

M31 has long been used as an example of what we would probably look like if viewed from afar: a colossal whirlpool-shaped object composed of orbiting stars. Our onrushing neighbor is even bigger, however: a galaxy of hundreds of billions of stars about twice the size of our own Milky Way.

The radial speed of M31's approach can be measured by the Doppler shift, or color shift, of the spectral lines of the light it gives off. While galaxies in the large-scale Universe are generally rushing away from one another, those in close proximity to one another often wind up on a collision course. It is clear that M31 is heading our way at a speed of over 300 kilometers a second, or 720,000 miles per hour. While presently at a comfortable distance of 2.2 million light-years (the closest star to our own is only about four light-years away) it nonetheless will get here in its own good time. Three billion years from now, long before the Sun becomes a red giant, it is expected that our Milky Way galaxy will collide and merge with M31, more commonly known as the Andromeda galaxy. This fate is not yet certain because we aren't sure the aim of our neighbor is accurate enough for a direct hit, but a smash-up appears likely. Even if the galaxies narrowly miss on their first pass, their mutual gravities will bring remarkable changes and an eventual merger. Like earthly corporations, the Milky Way and Andromeda will most likely combine to form a new entity. In all likelihood the Sun and its planetary bodies will remain as part of the merger, but perhaps not. There is a small but real chance that the gravitational forces of this encounter will not pull us in but cast us out, and our solar system will be flung outward into intergalactic space.

Why do astronomers anticipate this? Looking at other galaxies, we see abundant evidence for collisions and close encounters. One of the clues that these interactions take place is the presence of "tidal tails," or wisps of stars greatly extending outward from the main parts of their galaxies. In many cases the stars in these wisps are escaping off into intergalactic space. Although most stars are retained during a merger, some are lost, and there is always a chance our Sun could be one of the ejected.

The alternative is to be in the middle of things when the two galaxies intermingle. This collision, occurring at one-tenth of a percent of the speed of light, sounds pretty dramatic. But while this will change the night sky, it is by no means necessarily disastrous. Galaxies are almost entirely made of empty space, a sprinkling of stars separated from one another by vast distances, and because of this their collisions are mostly soft and gentle events that occur over hundreds of millions of years. The collisions are more like the intermixing of two groups of ballerinas than the crashes of trains, the belly bumps of sumo wrestlers, ships hitting icebergs, or asteroids hitting planets. Think of it as swarms of bees passing in the night. Over the past several decades, astronomers have recognized that collisions of galaxies are common and that this is just the normal way that nature does business. Massive galaxies often cannibalize their smaller neighbors. The Milky Way has probably assimilated many smaller galaxies, and, in fact, it is presently colliding with and assimilating the Sagittarius Dwarf Elliptical galaxy. Have you noticed?

Stars are so small, compared to the vast distances between them, that they essentially never physically make contact. They are like sparks in the wind, moving where the local forces of gravity move them. Incredibly, the gravity that controls the movement of the sparks is not due to the stars themselves—massive though they are—but instead it is primarily due to a mysterious and unseen material called dark matter.

Galaxies are much more than they seem to be. They are remarkably complex assemblies composed both of visible matter such as bright stars, and "invisible" matter. One of the greatest astronomical discoveries of our time has been the finding that galaxies are not just collections of stars mixed with a little gas and dust. The visible matter in images of galaxies represents only about 10 percent of their actual mass, like the visible tip of an iceberg is only a fraction of its underwater size. Most of the mass of galaxies is a mysterious medium called dark matter, because it is so dark our telescopes can't see it. Its presence is inferred by the effects of its gravity: it makes galaxies rotate, clump, and travel differently than if they consisted of stars alone. Each galaxy is embedded in a halo of this gravity-producing dark matter, but we don't know yet what it is. Ideas range from its being nonluminous planet or star mass objects referred to as MACHOs, or massive compact halo objects, to some unknown elemental particles generically referred to as WIMPs, or weakly interacting massive particles. Never let it be said that astronomers don't have a sense of humor.

In addition to dark matter, galaxies have other remarkable aspects. Most appear to have massive black holes in their centers, and our Milky Way has a black hole that appears to have a mass more that 2 million times as great as the Sun. Individual galaxies are also a sundry bunch that have remarkably diverse evolutionary histories and properties. Some are small, some are large, some are spiral-shaped, some irregular, and some are elliptical. Some galaxies produce stars at a continuous rate while others peak during the earlier or later parts of their histories. Like people, each galaxy is unique.

The spatial distribution of galaxies is also much more complex that one might expect. On local scales, galaxies are grouped into clusters. The measurement of distances to large numbers of galaxies has revealed that the Universe is a mix of regions densely populated with galaxies and huge voids where galaxies are rare. A slice through the Universe looks somewhat like Swiss cheese, or, better yet, a foam with large voids, surrounded by strings and sheetlike clumps of matter.

The very existence of galaxies, clusters of galaxies, and large-scale structures in the Universe has its roots back at the very beginning of the Big Bang. Slight variations when the Universe was smaller than a pinhead led to a distribution of mass that was clumpy. As the Universe expanded and cooled, the clumps of matter collapsed to form galaxies and clusters of galaxies on a grand scale. Without that lumpiness, matter would not have been able to form stars and galaxies, and we wouldn't be here.

Our Milky Way resides in a cluster of galaxies known by the chummy name the Local Group. Altogether there are more than thirty known members; most of the galaxies are small and some contain only a few hundred stars. (Most of the mass of the local group is in M31 and our Milky Way, each being a collection of more than 300 billion stars plus dark matter and a massive central black hole.)

The members of the Local Group move like a swarm of slowly moving jellyfish. Pulled together by mutual gravity but obeying the universal law of inertia, they follow chaotic orbits about one another. While the gravitational dance of two bodies is relatively simple and worked out by Isaac Newton some three hundred years ago, the motion of many bodies is chaotic and can be simulated only by computers. These simulations have shown that galaxies in clusters collide or make close passes to one another. As mentioned earlier, our Milky Way galaxy is currently colliding with the Sagittarius Dwarf Elliptical, a process that is causing the disruption of the dwarf and the eventual ingestion of its stars. Because our gravity is so much greater, the smaller galaxy will be torn asunder and most of its stars will become part of the Milky Way. There is nothing new about this appropriation. While most of the stars in our galaxy formed within it, some have been "stolen" from nearby galaxies during these close encounters.

For the approximately 13-billion-year history of the Universe, Andromeda and the Milky Way have largely managed to stay apart from each other. It now appears that they are on a collision course. Collisions can dramatically change the structure of galaxies, and the redistribution of gas can radically change the formation rate of new stars, increasing it so spectacularly that it's called a "starburst." The most massive stars in the starburst evolve very quickly and become supernovae that explode with incredible violence. The supernova rate in the Milky Way presently is about one per century, but in a starburst period it might increase to more than one a year. This dramatic increase in stellar explosions would be spectacular to future Earthlings if they had eyes to observe it, but as explained earlier, it is highly likely that by the time this accelerated star formation occurs, terrestrial life will have reverted back to a bacteria-only world.

The collision between the Milky Way and Andromeda has been simulated by John Dubinski at the University of Toronto and Lars Hernquist at the Harvard-Smithsonian Center for Astrophysics, using an IBM supercomputer calculating the motions of 90 million mass points, or "stars." The simulation is still not sophisticated enough to predict accurately the consequences of the collision in

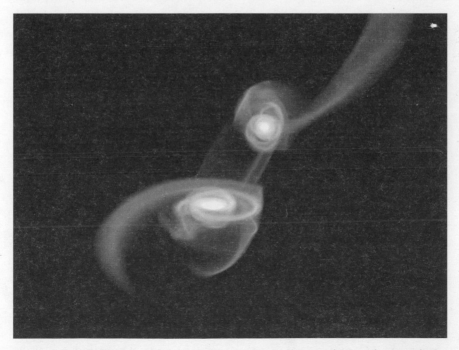

Computer simulation of a stage in the coming collision of our galaxy with the Andromeda galaxy. The image is courtesy of John Dubinski at the University of Toronto.

detail, because the properties of the two galaxies as well as their transverse speed are not accurately known. But they do show the general character of the spectacular and beautiful behavior associated with the collision of two large spiral galaxies.

The collision is a protracted affair taking more than a billion years. As the galaxies approach each other both develop spectacular spiral structures with wispy ghostlike arms extending outward. The spinning, fantastically distorted bodies essentially pass through each other while the mutual gravities from their stars and dark matter act on each other. The galaxies actually pass through each other, like bull and matador, but then fall back together. Most of the masses of the two galaxies ultimately combine to form a new galaxy. A small fraction of each is lost in escape trajectories that take them into the loneliness of intergalactic space.

Assuming that the Earth stays with the galaxy, it will find itself in a different galactic environment. Once the new merger is completed, our solar system will reside in an elliptical galaxy devoid of the Milky Way's spiral structure. Stellar orbits become both less circular and more inclined (to each other). After an initial burst of new star formation, the gas that fuels it declines and the rate of formation of new stars drops drastically. As an elliptical galaxy, there are no spiral arms marked by new stars and star-forming regions, similar to the Orion nebula in our present sky. With its birthrate low, the entire galaxy becomes an aging structure of ancient stars slowly fading into oblivion.

If Earthlings could watch the whole collision process as a highly accelerated time-lapse movie, they would basically see M31 get larger and brighter until it eventually covered the whole sky like a second Milky Way band. After the initial approach, the event would probably not be very spectacular with the exception of the enhanced rate of supernova explosions. The galaxy's changing shape would not be apparent from the Earth's vantage point. Trying to see the event would be somewhat akin to trying to admire the intricate structure of a corn maze or crop circle from ground level. Large-scale structure is best seen from afar.

Complicating the issue is the presence of dust. Vast numbers of tiny particles of stone and carbon already block our present view of our galaxy. When you look at the Milky Way right now you are seeing the inside, edge-on view of our own disk-shaped galaxy. The diffuse glow of unresolved stars arching across the sky has prominent irregular dark bands along the centerline, which are regions where dust is blocking out the light of distant stars. Essentially all of the individual stars that we see in the night sky are local, both because closer ones look brighter and because more distant ones are obscured by thousands of light-years of interstellar haze. This haze will also obscure the grand collision. From Earth, it may not differ substantially from the present view seen from the Southern Hemisphere where both the Milky Way and the Magellanic Cloud galaxies appear as a complex pattern of diffuse glows.

After a billion years of interactions, the merger will be complete, things will settle down, and the new elliptical galaxy will appear as a brighter, broader "Milky Way" without dust lanes because the dust will largely have been consumed by new stars. This will be a more regular, more diffuse, and more mundane sky, a sky without M31. And this is the sky that will dominate the night when the last creatures on Earth are making their last stand against the relentless increase in brightness of the Sun and increasingly hostile conditions on Earth.

BILLIONS OF YEARS HAVE PASSED SINCE THE COLLISION WITH M31, billions of years in which the Sun has steadily grown brighter. Perhaps this enhanced warmth has allowed life to begin a different trajectory on the moons of Jupiter or Saturn. Certainly Mars is now warm, instead of frigid. But on Earth, the time of life is long gone.

Step with us onto this seemingly alien planet of the future.

It is incredibly hostile. The atmosphere is dense and hot, like that of present-day Venus. Thick clouds of sulfuric acid and residual water enshroud the planet and we cannot see the gigantic Sun, but its presence, its heat, is unmistakable. The day is gray, without shadows, and dark nights are a thing of the past. The Moon, which had been retreating outward for billions of years, is beginning to spiral in, on its last ride before its ultimate collision with the Earth. It has been 11 billion years since the origin of the Earth, and the Sun has entered its red giant phase. Its tendrils begin to sear the Earth.

As menacing as this swollen Sun is, we should be grateful for its long service. Before we consider Earth's final days, we should consider how short our planet's history might have been if the Sun had had a slightly higher mass. Almost paradoxically, if the Sun were more massive, its lifetime would have been shorter. More massive stars are hotter in their centers, particles fly and fuse faster, their nuclear reactions occur at a faster rate, and they correspondingly run out of fuel faster. If the Sun had 50 percent more mass—not at all unusual for stars—it would have already run out of hydrogen fuel and become a red giant cooking our planet. Because of this accelerated history of a more massive Sun, there wouldn't have been time for animals to flourish. The Sun that we enjoy today gives animals about 1 billion years to evolve, a period we are just over halfway through.

Conversely, if the Sun had been smaller, there would have been less heat, creating a myriad of related problems. Stars range from ten times smaller to nearly a hundred times bigger than the mass of our Sun. It is only chance that our star fell into the right range.

Although the understanding of the evolution of stars required a century of scientific detective work, we actually see wonderful examples of the different stages of stellar life every time we gaze at the night sky. In a few moments of observation we can see examples of the whole sequence of stellar evolution. A good place to start is Orion (the hunter), a well-known constellation that dominates the winter sky. If we use binoculars or a small telescope to look at the sword hanging from Orion's belt, we see a glowing patch containing many bluish stars, the famous Orion nebula. This is a stellar nursery only three hundred light-years distant, where new stars are forming at the present time. The brilliant blue stars are less than a million years old, and their glow of ultraviolet and visual radiation lights the surrounding envelope of gas and dust to produce the spectacular nebula. In a single glance we can see direct evidence of the birth of stars from interstellar gas and dust.

If we look at the right shoulder of Orion we see a bright red star that is a dazzling example of the near-final stage of the life of stars. Marking Orion's shoulder is Betelgeuse, a supergiant star that is in the last few million years of its life as a bright object. It is spectacular, more than ten times as massive as the Sun and more than fifty thousand times as bright. If it was placed at the center of the solar system, its surface would reach halfway to Jupiter, and the intense glow of its reddish light would produce incinerating temperatures on all of the planets, even frigid Pluto. As we gaze at Betelgeuse we are viewing the death of a star. Its nuclear processes are near the end, and the star has not only swollen to gargantuan size, it is unstable, varying in size and brightness over periods of only a few years. This star is approaching the most dramatic stage that a star can experience: Betelgeuse will soon become a supernova. It will explode, and for a period of a few days it will generate as much light as the entire Milky Way galaxy.

Our own Sun will never reach this fate because the grandest exit for stars is reserved only for ones, like Betelgeuse, that are at least eight times more massive than the Sun. Our Sun will become a red giant, but it will never become a supernova.

Among the bright stars of the night sky we can also see stars in the same evolutionary state as the Sun. Sirius (also known as Canis Major, or the dog star) is the brightest star in the night sky. It is seen as a brilliant blue star often scintillating with a rainbow of colors due to the effects of our turbulent atmosphere. It is a main-sequence star in its middle age, doing what middle-age stars do:

generating energy by burning hydrogen into helium. Because it is so massive, Sirius will run out of nuclear fuel much faster than the Sun and eventually become a red giant somewhat similar to Betelgeuse.

But not exactly similar. Unseen by the naked eye is the fact that Sirius is a binary, or double, star. It has a remarkable companion that orbits around it every fifty years. This partner is too faint to be seen with the naked eye, but was first seen with a telescope in 1862. The second star is slightly more massive than the Sun but smaller than the Earth! Sirius B is the first white dwarf star ever seen. This star was originally more massive than Sirius, but it quickly burned up its nuclear fuel and exploded, ejecting most of its mass into space. The remaining fraction survived as a tiny, incredibly dense, and rapidly spinning white dwarf. White dwarfs are the terminal stage of stellar life, when stars have been transformed into highly compact bodies that slowly cool to oblivion. The Universe is littered with white dwarfs. They are difficult to see because of their faintness, but they are actually among the most common types of star. Eventually *all* the stars will be either white dwarfs, neutron stars, or black holes: the evolutionary end point.

THE SUN OF OUR OWN TIME IS IN MIDDLE AGE, AN EVOLUTIONARY stage that occupies about 90 percent of its lifetime as a bright star. As it depletes the hydrogen fuel in its interior it gradually becomes brighter, and during its 11-billion-year main-sequence lifetime, it will brighten about two and a half times. The changeover from a main-sequence star—yellow and normal-sized—to a red giant will be dramatic and short. Its helium core burns ever hotter, pushing out its surface shell of gas. This paradoxically cools the surface, making it appear redder in color. The diameter swells, and our star becomes thousands of times brighter than it is at present. If a person could watch from Earth, he or she would see the Sun expand from its present angular diameter of half a degree to a star that completely fills the daytime sky. Its radiation power increases two thousand times, even as its surface temperature drops to half its present value. Even without greenhouse gases, the surface of the Earth will be heated to over 2000 degrees C, enough to melt mountains and turn our planet into a billiard ball.

As the helium-rich core becomes ever more massive and dense, its atoms enter a most remarkable state, becoming so crowded together that they form what is known as "degenerate matter." In degenerate matter, pressure is not determined by temperature and so the Sun cannot remain stable by expanding and cooling. Instead, the core becomes explosively unstable. The result is a helium flash.

When the core temperature of an evolving red giant reaches 100 billion degrees and the density is a million times greater than the density of water, the helium core explodes. In these extreme conditions helium fuses to form carbon, releasing extraordinary amounts of energy in a very short period to time. This occurs by what is called the "triple alpha reaction." Three helium nuclei (alpha particles) essentially collide at the same time and fuse together to form carbon. This is one of the most important nuclear processes because it is the gateway for producing heavier and heavier elements. Carbon is the fourth most abundant element in the Universe, and essentially all of it was produced by the triple alpha process. It requires the high temperatures and densities that occur only late in the life of a star, and it is this kind of flash that forged the elements required to make you, given that you are a carbon-based life-form.

In a helium flash, several percent of all of the helium in the core is turned into carbon in just a few minutes. When the core begins to flash, heat is generated, but the core cannot expand to cool itself and stop, or even slow, the reaction. The hotter the core, the faster the reaction goes, and for a few minutes the energy generation of the core of the Sun increases by a factor of 100 billion times its present rate. This is the energy equivalent of a supernova explosion—but none of the energy gets out! Instead, it goes into converting degenerate matter into normal matter, such as carbon. The star then undergoes a remarkable transformation, ejecting some of its mass into space, dropping in brightness by a factor of ten, and becoming ten times smaller than it was previously—though it will still be ten times larger than its present diameter. After the helium flash the Sun enters a 200-million-year time period when it is fairly mundane, burning quietly.

The stage is then set for the Sun's final days as a luminous star. It enters what is known as the Asymptotic Giant Branch or AGB stage, the stage that will destroy the solar system's inner planets. The Sun becomes unstable and undergoes violent thermal pulses where its size and energy generation surge to unprecedented levels. The Sun's diameter pulses outward to the original size of Earth's orbit, and it becomes as much as six thousand times as bright as it is at present, hot enough to melt the surface of the Earth. As the Sun flares out, its atmosphere produces a drag force on the Moon, allowing it to spiral inward and return to Earth with colossal impact. The irony is that this completes a cycle. The Moon formed as the result of a giant impact between Earth and another planet early in the history of the solar system. Once ejected, it now returns.

Yet even this isn't the final doom. As the Sun's diameter swells, it absorbs the inner planets one by one. First Mercury is vaporized. Then Venus. And Earth? It is possible it might just escape and survive as a cinder. Our planet is at the orbital distance from the Sun where planets can either survive or be swallowed. While current calculations suggest the Earth will be consumed, the details of the Sun's final evolution are not known well enough to confidently rule out a near escape. During its red giant phase, the Sun loses approximately half its mass to space. As its gravity lessens, all of the planets move outward, which means the Earth might narrowly survive as a planet. On the other hand, tidal interaction between the Earth and the Sun could cause our planet conversely to spiral inward, dooming it after all. This is the destiny predicted by astronomers Kaeper Rybicki and Carlo Denis in a 2001 paper in the journal *Icarus*.

If the Earth is absorbed, the final end will be by fire. Slow orbital decay and expansion of the Sun during one of its thermal pulses will bring our planet into the actual atmosphere of the Sun. The Earth will fly at over 20 kilometers per second through the hot gases, becoming incandescent with a spectacular supersonic bow shock. The drag of the Sun's atmosphere would cause the planet to spiral deeper and deeper until it would be entirely vaporized at temperatures above 100,000 degrees. Every chemical bond in the planet and every shred of information about its past state would be lost. Twelve billion years of planetary history will be reduced to individual atoms, many of them hurled back into space as the Sun expends its gas. In the distant future, they may be incorporated into new solar systems, planets, and perhaps life.

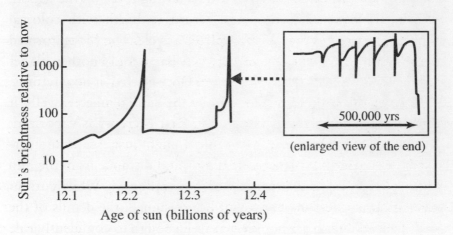

After 11 billion years of gradually increasing brightness, the Sun becomes a red giant, a series of phases when it becomes vastly brighter, much larger, and ejects matter into space. The peak, just after 12.2 billion years, is where the remarkable "triple alpha reaction" occurs. The energy generation rate briefly increases 10 billion times but the brightness of the Sun, its energy radiated to space, paradoxically plummets. The pulses near the end are shell flashes within the star that cause the greatly expanded Sun to spectacularly eject matter into space, forming a planetary nebula. The last and brightest flash may be the one that finally envelops the Earth, causing its complete destruction.

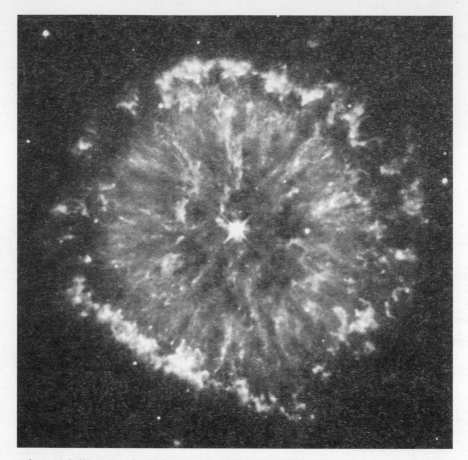

After 12 billion years of life the Sun will eject nearly half of its mass into space in a series of brief but spectacular bursts of high-speed gas. This *Hubble Space Telescope* image of the planetary nebula NGC 6751 shows a star 6,500 light-years away that is in the final explosive stages of its life, a stage the Sun will reach in about 7 billion years. The expanding gas cloud is six hundred times larger than the orbit of Pluto and it can actually be seen with a small backyard telescope. It is a vision of our future.

In its final stages the Sun has a few last-minute tricks. During several of its thermal pulses, it will release large amounts of matter into space that will form what is known as a "planetary nebula." These luminous gas clouds surrounding dying stars have nothing to do with planets, but they do provide some of the most spectacular images in astronomy, forming fantastic clouds of luminescent gas rapidly escaping the star that ejected them. If life on Earth had managed to escape to some distant sanctuary of the solar system, it would probably be finished off by the intense radiation of these events. Even if Earthlings had, in final desperation, fled to a solar-orbiting colony beyond the orbit of Pluto, they would still be annihilated by the final days of furor of their once benevolent Sun.

After the Sun's giant phase, it rapidly settles down to become a white dwarf. Its brightness declines thousands of times and it shrinks to only 1 percent of its present diameter. With its nuclear fuel mostly expended, the white dwarf Sun will become dimmer and dimmer with time, but it will faintly radiate for billions of years. As viewed from Mars, now the closest surviving planet to the Sun, it will glow with the brightness of the full Moon. It will be a silvery but cold glow, with no life-supporting heat. This will be the final state of the Sun, a ghostly marker.

This is the normal fate of solar systems, as inevitable as our own individual deaths. The galaxy contains billions of white dwarfs, many of them just tombstones for worlds like our own.

# · 10 ·

# ACCIDENTAL ARMAGEDDON

THE POET T. S. ELIOT WROTE THAT THE WORLD ENDS NOT WITH a bang but with a whimper, but popular culture desires a more catastrophic end. The fiery consumption of Earth by a swollen Sun is spectacular enough, but astronomers forecast it won't take place for another 7 billion years or so: a time somewhat distant to lie awake worrying about. Aren't there possible catastrophic ends that could happen much sooner: such as tomorrow?

We've all seen this movie: here is the world, going about its business, and *blam*—out of the blue (or starry black) comes a fire-and-brimstone cosmic catastrophe of some sort that quickly ends it all. Armageddon! If our scientific predictive powers are at all correct, an accidental end to the world is the *least* probable of the ends that we can envision. Yet the chance exists. Could it really end the world?

The most common scenario is nuclear war, and Armageddon is quite familiar to anyone who lived through the most paranoid part of the cold war. The Cuban Missile Crisis was more than a movie. The Berlin Blockade was real. Both authors grew up in the 1950s, and that upbringing marked us. Once a week a siren would go off, testing the civil warning system designed to tell us that within twenty minutes a hydrogen bomb would be coming to our very neighborhood, and that it would kill us unless we could find shelter in some neighbor's bomb shelter, or at least underneath the desk at school. Sudden catastrophe was thus as much a part of our childhoods as baseball and summer vacation. Sometimes the sirens would go off for no reason, and everyone would twitch and look up at the sky. Just to make sure we could visualize this eventuality, Hollywood and the huge science fiction industry kept up a constant parade of movies and books about nuclear Armageddon and its ghastly effects. Who of that time period can forget Nevil Shute's *On the Beach*?

While the cold war has ended, additional nations continue to acquire the bomb, and terrorists long for one. The threat has not gone away. Today, it is probably more likely that a nuclear war would be limited rather than total. Or that even if human civilization were hurt or crushed, our species would survive in more primitive fashion. Or, that even if *Homo sapiens* erased themselves, other creatures would inherit the Earth. Still scientists have proposed nuclear-winter scenarios that could cause mass extinctions, and the seriousness of this threat should never be underestimated. Even so, thinking the unthinkable in nuclear terms is beyond the scope of this book. So are science fiction or religious scenarios such as global plagues, ecosystem collapses from overpopulation, a takeover by robots or computers, alien invasions, or a Second Coming. It is not that such things are impossible, it is simply that they are beyond the predictive powers of physical science.

There have been real Armageddons in Earth's past, however, caused by the kind of asteroid or comet impacts that were common in the planet's early history. Thankfully, these have become extremely rare. Yet although most of the early solar system debris that crossed Earth's orbit is gone, some remains. Most are dust-sized and are called Brownlee Particles in honor of you know who. Our other coauthor has spent part of his career investigating what happened to the Earth when much larger particles struck home. Sixty-five million years ago, an asteroid perhaps as much as ten kilometers in diameter hit the Earth in the Yucatán Peninsula in what is known as the K/T impact. It is calculated that a K/T-sized rock hits the Earth every 100 million years, but this is an average, not a count-down. We could have 35 million years or more until the next one. Or it could happen at any time.

Let's imagine that a similar-sized body hits again, in the same spot in the Gulf of Mexico, in our time. What might that impact look like if viewed from Texas and Colorado?


THE HERD OF COWS LOWED TO ONE ANOTHER AS THE FIRST hints of dawn began to paint the eastern sky. The low coastal pastureland east of Houston was already alive with the calls of birds, but calls somehow dissonant and anxious, as if predators were approaching—yet none could be seen or smelled. The only visible peculiarity was to the south where a bright star blazed forth, its illuminated tail extending across half the night sky. This star had been growing in size for many nights, growing brighter with each nightly reappearance. The birds and insects that depended on the Moon for navigation were becoming distracted and troubled. Seaward, in the warm Gulf of Mexico, the new celestial light gave extra light for fish to hunt for food.

The cattle herd was completely awake now and starting to feed on new grass. A pink glow of dawn was accompanied by swarms of insects rising, many settling on browsing cows and sheep, others flitting through the humid atmosphere. Overhead, the long glowing tail of the great comet began to disappear as the night sky was over-taken by dawn, but its disappearance was as much a matter of its final approach as it was being overpowered by the approaching day. The bright head of the comet could be seen slowly descending to the south, finally disappearing below the horizon; it was then followed seconds later by the orange glow of a second dawn.

A brilliant bar of white light shot upward into the sky, proclaiming the end of one era and the beginning of a new one. Molten rock from the impact created this seeming beam of light, rock from the comet and the impact site intermixed as one, and it blasted into the thin pillar of vacuum created by the comet's fall to earth. The cattle were still quite oblivious, southern light shows being irrelevant to their dull, domesticated brains. But the birds paid heed, and their usual dawn cacophony fell silent as this second dawn unfolded. The thin pencil of light began to change color, become more diffuse, and widen; from its base tiny specks of light fanned outward in all directions. A clap of what sounded like distant thunder now startled the cows as well, and all turned with frightened eyes to the south as a low bass rumble began to amplify. A low cloud appeared, moving rapidly toward the herd. The cows began to turn in terror as the shock wave approached and then it hit with speeding seismic fury, emptying the trees of birds and knocking all fleeing mammals to their knees or onto their backs. Then it was past, the cattle staggering up and beginning a mad rush in all directions as an orange cloud crept upward from the southern horizon.

The first of the meteors began to streak overhead, buzzing like mad hornets. More followed, screaming inward, hitting the sea some miles from shore with loud explosions. Others passed overhead or fell earthward like artillery shells, crashing with increasing frequency, and still the sky filled with more shooting stars. The Sun, rising at last in the east, was overpowered by the streaking light show and then obscured by the rising walls of smoke, for the superheated bits of rock were beginning to set the Texas coastal forests and grasslands alight. The fires were isolated at first, still contained by the wetness of the coastal plain. Gradually, however, they began to link up as the air and forests heated, and as more and more stars streaked inward, gravity's rainbows.

The sky overhead was now brilliant yellow, a daylight of shooting stars, the Yucatán sea bottom intermingled with celestial debris. The wind rose into a firestorm. Within the first few hours the vegetated regions became sterilized, unnumbered creatures great and small exterminated by the fire and heavenly brimstone.

In the nearby sea the shooting stars pummeled the shallows, raising the temperature of the top meters of ocean toward a boil. Fish began swimming downward toward deeper, colder water, but escape became irrelevant as a monstrous current tugged all of the ocean denizens toward the south. In this place, no depth was deep enough to be a safe haven. In the muddy sea bottom, burrowing species that for millions of years had lived peacefully enough in their muddy homes became involuntary intertidal creatures for the first time, gasping or writhing in this first exposure to air. The sea was being sucked away, and the bottom a hundred feet below the normal ocean surface was suddenly exposed by the rapidly retreating sea.

Along the shorelines, terrorized animals and people followed the sea's retreat, fleeing from the blazing forest into the now exposed mud and sand, the only place not burning. And still the sea retreated. It was a world gone mad, a world gone topsy-turvy: the green cool forests now aflame, the once calm sea racing away from its shores, the sky ablaze with sheets of shooting stars, and the landscape whipped by superheated raging winds carrying with them the screams of burning creatures. Inland, swamps and lakes became the last refuges for animals and humans alike. Survivors crowded the shorelines, shell-shocked by the cosmic barrage, explosions blasting great gouts of burning forest into the sky.

Then the sea returned.

It came with a great roaring, a black mountain of water a kilometer high, reflecting back the sparkling red of the burning land. The Caribbean and Gulf of Mexico had been pushed from the Ground Zero crater into the largest wave the world had seen since the K/T catastrophe of 65 million years before, which finished the dinosaurs.

The huge tsunami smashed Texas cities into oblivion: first Houston, then San Antonio, then Austin, and then distant Dallas. Still the incoming meteors bombarded the Earth, splashing into the giant brown sea littered with trees, boats, fragments of houses, and millions of bodies. Eventually the great waves subsided, leaving behind high-water marks of death. In Texas, nothing survived except the fish and bacteria.

Far to the north and west, near Denver, the dawn broke more normally—but soon the first shooting stars began to fill the sky. Jets trying to land at Denver International Airport were forced into panicked evasive maneuvers as disbelieving pilots watched meteors fall like flaming rain. Some were hit and some not, but success was moot as the Denver airport runways rapidly became littered, causing pilot after pilot to set down in the surrounding fields or on freeways already filling with panicked motorists. Sometimes they landed successfully, sometime as fireballs, but it mattered not. Most escaping passengers were soon struck down by the barrage of rocks from the sky.

The Great Plains caught fire. Grassfires began in a thousand places until a solid wave of flame blackened all of the grass and then died out as it ran out of fuel. The clear morning sky began to darken as a great black cloud arose from the south and merged with the smoke from the now-burned fields, a blackness of spreading dust and smoke tinged by lightning and illuminated by the spitting, flaming rocks that still fell unceasingly. By midmorning the Sun was totally obscured by the dust cloud, and with the new night at noon came dirty rain, a rain that grew colder and increasingly acidic as the days passed. It was a rain that scalded the skin of the survivors, ate away the protective surfaces of their clothes or fur, and poisoned plants in the ever-deepening cold. The Sun would not shine again on the Denver plain for a year, and when it did, it was on a skeletal landscape of trees killed by summer frost, in a place were summer was banished for years to come.

The Sun finally rose again on the bleaching bones of the human world. Here and there life still existed. Cockroaches, beetles, weeds, and many, many rats alongside the few surviving and starving humans. They envied the dead.

WOULD SUCH A SCENARIO QUALIFY AS AN END OF THE WORLD? The last time this played out, as many as two-thirds of all species on Earth went extinct. For the dinosaurs it was the end. But much survived and the Age of Mammals emerged from the ashes.

How likely is this scenario? In 1995, a great comet was discovered simultaneously by Alan Hale and Thomas Bopp, using amateur telescopes. The comet was 40 kilometers in diameter, and it crossed the orbit of Earth. Had it collided with our planet, it would have delivered nearly a hundred times as much energy as the comet or asteroid that killed off the dinosaurs, and it would have caused extinction of most species of plants and animals. Yet even comet Hale-Bopp would not have ended all life. For that to happen we'd need a comet or asteroid two or three times bigger still.

Comets like Hale-Bopp come in from the far reaches of the solar system. When they are discovered by telescopes, they are usually just months from Earth's vicinity. Even with the advanced techniques and capabilities of future humans, it is unlikely that Earthlings would have the time, patience, and preparation for a successful defense mission, even though such rescue missions have been portrayed in movies. Even in the future, Earthlings might be helpless, for there is no telling how long we will be able or willing as a society to produce large rockets capable of interplanetary travel and capable of delivering hydrogen-bomb warheads. Because such impacts are so rare, it is difficult to imagine human organizations being successful with such long-range planning and allocation of needed funds or resources necessary to deflect a truly large comet from hitting Earth. Who would get ready for something that might happen soon but probably won't happen for millions of years, and perhaps not happen at all?

Yet what if it did? What would a truly gigantic comet or asteroid—something 150 kilometers in diameter—do to Earth?

It would streak in at 40 kilometers a second. At that speed and size, it would generate so much heat and energy that if it struck the Pacific, most of the ocean would vaporize and turn to steam in minutes. The vaporization would not be instantaneous—the whole process would actually take some minutes—but when it was over, water as a liquid would no longer exist in Earth's biggest ocean. The impact would also blow a significant proportion of the Earth atmosphere into space but that's irrelevant, for the flash of steam would cook the entire surface at several hundred degrees. The planet would be sterilized. In a brief period of time, every single organism on planet Earth would be dead.

THE IMPACT FROM AN ASTEROID OR COMET IS NOT THE ONLY quick end to life on Earth that we can envision. Energetic bursts emitted from exploding stars are also a potential threat. Like the description of the effects of a truly large comet on the Earth, the story of how a nearby gamma ray burst might affect life is short. One minute you exist, and the next you are either dead or dying from radiation poisoning.

It has long been suggested that radiation blasts from exploding stars in our galaxy might cause global mass extinctions, but so far there has been no definitive evidence that such an event has ever happened. The largest radiation threat is probably from gamma ray bursts, some of the most remarkable and enigmatic processes that have been detected by space telescopes. They are the most powerful of all natural events. About three times a day the Earth is impacted by bursts of gamma rays, or high-energy photons, that come from random spots in the sky. The pulses last from less than a second to minutes.

These pulses were first detected in the 1960s by gamma ray detectors put into Earth orbit to monitor possible violations of the Nuclear Test Ban Treaty. Instead of finding clandestine nuclear tests, the satellites detected mysterious and intense bursts of gamma ray energy that are still not fully understood. They come from random directions, most from beyond our own Milky Way galaxy. Because of the shortness of the bursts, only a few seconds, their sources must be small: only a few light-seconds across, similar to the distance between the Earth and Moon. Yet some of the bursts come from the far reaches of the Universe and because of the long distances traveled and their incredible intensity at Earth, their source, however small, must be exceedingly energetic.

If you could see gamma rays with your eyes, you would see the sky light up about once a night, but these distant events go unnoticed by our natural senses. If one ever happened close by, however, it would be like the difference between watching lightning flashes in a very distant storm cloud and being in the actual path of a million-amp thunderbolt.

The origin of gamma ray bursts is still a matter of debate, but a major theory is that they are associated with the supernova explosions of very massive stars. When a massive star reaches the end of its life, it blows up. Its core will form a black hole, and it is believed that associated processes will send two opposing beams of intense radiation off into space. Like an extreme science fiction tale, these beams are death rays. Such a ray in our own galaxy could pierce much of its length, terminating life in millions of solar systems unlucky enough to be in the beam's path.

How likely is such an event? Presumably rare, since Earth has not been irradiated to extinction. Yet not impossible. In the year 1054 a most amazing event occurred. In that year a star relatively near our own Sun became a supernova and was so bright that it could be seen in the daytime. Today, the remnants of this cosmic catastrophe are still visible to anyone with access to even a modest telescope. The cosmic debris left over from this event is now known as the Crab nebula, a shredded cloud of gas that less than a thousand years ago was a star.

European astronomers didn't record the Crab nebula supernova, but they were definitely present and observing for the next great supernova explosion, in 1572, and then another in 1604. Since that time there have been no other supernovas in our Milky Way galaxy, but modern astronomical instruments have allowed us to observe the effects of supernovas taking place in other galaxies. Eventually there will be a supernova in our own galaxy powerful enough to cause a gamma ray burst.

The star Eta Carinae is the most massive star known in the Milky Way, and it will surely explode as a supernova within the next million years, producing a gamma ray burst. Eta Carinae is a hundred times more massive, and 5 million times brighter, than our own Sun. Even though it is seven thousand light-years away (our nearest stellar neighbors are only about four light-years away) its gamma ray burst could be powerful enough to kill us with radiation and destroy our planet's protective ozone layer. Fortunately, gamma ray bursts are shot from a star's poles, and Eta Carinae's poles are not pointed at us. But are there other stars that are?

Who knows? It has been estimated than any gamma ray burst within twenty-five thousand light-years would be sufficiently close to cause global mass extinction, but astronomers have not done a census of stars to determine which ones might blow and whether they are pointed in the wrong direction. In any event, we would have no warning of this extinction. The gamma ray burst travels at the speed of light, the same speed of light coming from the exploding star, and both would reach us at the same time. In the blink of an eye, all life on the planet could wink out. Fortunately, this is obviously an unlikely occurrence.

So what are our chances? There is no way of telling. There is no cosmic life insurance, or stellar bomb shelter. We are at the mercy of chance amid the cosmos. If there is any good news, it is that while the kind of asteroid impact depicted in the movie *Deep*

*Impact* might destroy civilization, or even our own species, it would not necessarily destroy all life on the planet. Earth would abide, and perhaps a new group of animals would triumph, as mammals triumphed after the dinosaurs. Still, we are a species that naturally longs for some kind of immortality. Would anything of ourselves be left behind?

· 11 ·

# WHAT TRACE
# WILL WE LEAVE?

IT IS CONTRARY TO THE HUMAN SPIRIT NOT TO HOPE THAT WE can leave some sort of enduring legacy. The end of our physical lives is often softened and rationalized by the belief in a spiritual afterlife. We have a drive to propagate and to preserve our own species. Although most of the species of higher animals on Earth have become extinct in only a few million years, we expect more for ourselves. We are so clearly different from all other animal species that there is an expectation that somehow, some way, we may survive. And if we can't achieve species immortality, we yearn for some kind of legacy denoting that we passed this way.

For millennia travelers have left graffiti to note their passage. Lewis and Clark and Daniel Boone carved their names on trees. The pharaohs left the Pyramids. Paleolithic man left magnificent charcoal drawings in Pyrenees caves. Mountaineers leave notes at the top of the peaks they conquer. World War II GI's left "Kilroy was here" scrawls, and the *Apollo* astronauts left plaques and flags on the Moon. What can we leave behind on an Earth that is going to be vaporized, its atoms hurled into space? Are there any records of us that will last? Is there any hope at all that we might be able actually to propagate our species beyond the life of the planet?

Curiously, we have already left an electronic legacy.

Even if the Earth is totally lost, there will still be faint, lingering evidence of our existence because our television, radio, cell phone, and data transmission signals are leaking into space. *I Love Lucy* has now propagated to a distance of more than forty light-years from Earth, and in principle it could be detected by advanced civilizations that might live around any of the nearest thousand stars. The possibility of our radio emissions providing a time-ordered "fossil record" was wittily pointed out in Carl Sagan's novel *Contact,* when the first Earth signal received from our alien neighbors was a television rebroadcast from the 1936 Berlin Olympics, making Adolf Hitler our first interplanetary emissary.

Since 1920 when station KDKA in Pittsburgh, Pennsylvania, began the first commercial transmissions, Earth has been broadcasting its news, culture, and sports, quite literally, to the four corners of the Universe. If a technological civilization existed on Mars, they would surely know all about us. Aliens beyond the nearest stars would have a much more difficult time eavesdropping because signal strength rapidly declines with distance and noise from other sources blankets them. Our broadcasts will travel forever, but they are so weak they will be undetectable after a thousand light-years at most, and perhaps much less.

If listeners are not relatively nearby, then this means of leaving our mark is probably futile. If we could find an audience, however, perhaps the best that could be hoped for to continue our memory would be to beam them our history and knowledge as a planetary time capsule. Even if we could not make physical contact because of the immense distances involved, a signal sent from our largest radio telescope could be detected by a similar device anywhere in the galaxy, if the dishes were directly pointed toward each other.

In addition to radio waves, we also have sent out physical records that will last extraordinarily long periods of time. Solid materials protected from heat, radiation, and interstellar dust can preserve information almost indefinitely. After all, we receive meteorites from the asteroid belt that retain excellent records of processes that occurred 4.5 billion years ago, a third of the age of the Universe. Similarly, if you scratch HI MOM on a piece of glass and place it a few meters below the surface of the Moon, it will remain unchanged until the Sun becomes a red giant. This type of longevity is quite extraordinary by Earth standards, where weathering and subduction destroy materials on much shorter timescales. The Earth contains very few rocks older than 2 billion years, and most of the ones that have survived have been strongly processed. Continental crust is constantly eroded and recycled, and within 100 million years—a blink in time in Earth's long history—the Himalayas, the Andes, the Sierra Nevadas, and the Rocky Mountains will all be worn away. The durability of man-made monuments is even more pathetic. The Pyramids, the Great Wall, dams, or freeways may last ten thousand years, but none will survive for the millions of years of geologic and astronomical time.

Accordingly, our best chance of preserving a record of ourselves is in places other than Earth. One of the first truly long-lasting time capsules from our civilization was the material left on the Moon by the Apollo program. The site of the *Apollo 11* landing contains a famous stainless steel plaque and drawings of both hemispheres of Earth, along with the statement, "Here men from the planet Earth first set foot on the Moon. We came in peace for all mankind." It includes the signatures of Neil Armstrong, Buzz Aldrin, Michael Collins, and Richard Nixon. While the plaque will slowly be degraded by the impacts of micrometeorites, it will be readable for a million years.

The six Apollo landers plus the equipment that they carried will survive for impressively long times. This equipment includes the Lunar Rovers carried on the last three missions, lunar surface experiments, digging tools, drills, a golf ball, and the plutonium power sources that powered the scientific measurements stations left on the lunar surface. These remnants will still be recognizable after a billion years. In fact, the landing spots will still be littered with identifiable debris at the time the Sun becomes a red giant, 5 or 6 billion years from now. The sites will resemble long-abandoned picnic grounds of long-forgotten but most untidy visitors. The ground will be littered by tens of thousands of impact-generated fragments of aluminum, copper wire, steel bolts, plastic sheets, and 1960s electronics. For billions of years the exotic materials of Earth and our civilization will sparkle in the sunlight against the mixture of fine gray powder, pebbles, and rocks that make up the lunar surface.

Compared to Earth, the Moon is an excellent place for storing long-range records because it has no eroding water, wind, or volcanoes. Still, even it will be destroyed by the expansion of the Sun. Truly long-term survival of Earth's physical records are practical only on spacecraft that leave the solar system.

We have launched several spacecraft on escape trajectories to interstellar space. They carry with them what will be the longest-lasting memory of our civilization to date because they will escape the impacts of comet and asteroid debris and all effects from the evolving Sun. They are not indestructible, of course: they will still be exposed to the impacts of tiny interstellar grains and their surfaces will be slowly sputtered away by interstellar gas atoms. Still, their lifetimes will be far longer that anything stored in solar orbit or placed on the surface of any solar system body. Compared to the lunar relics of the Apollo program, the solar system escapees will essentially last forever. Several of these spacecraft have carried messages. While the likelihood of the messages ever being found and read by alien beings seems quite remote, they represent our first steps toward providing an enduring legacy.

These message bearers include the *Pioneer 10* and *Pioneer 11* spacecraft, launched in 1972 and 1973, which have already left the solar system after visiting nearby planets. *Pioneer 10* is headed in the direction of the constellation Taurus, and *Pioneer 11* is headed

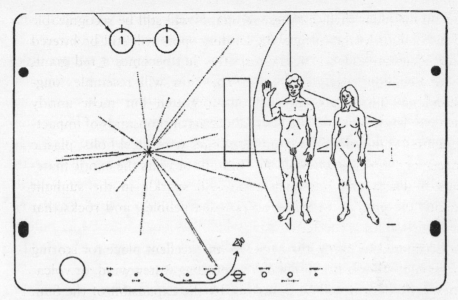

This metal plaque carried on the 1973 *Pioneer 10* spacecraft was the first physical message beyond the solar system. Conceived by Carl Sagan and Frank Drake, it spawned more elaborate messages carried on *Voyager* and later missions. These messages are likely to be the longest-surviving artifacts of our civilization.

toward Aquila. Within a million years, both spacecraft will be closer to other stars than they are to the Sun. Both carry identical six-by-nine-inch gold-anodized aluminum plaques as postcards from Earth. Engraved into the plaques were line drawings of a naked man and woman, the spacecraft, the solar system, the hydrogen atom (how important it has been to this story!), and the positions of nearby pulsars that might be used to determine the port of origin.

The *Voyager 1* and *Voyager 2* missions were launched in 1977, and, after a stunning tour of Jupiter, Saturn, Neptune, and Uranus, they followed the *Pioneers* into interstellar space. They bear highly sophisticated and well-thought-out messages that, if ever found, could provide a wealth of information about our civilization: 115 images, greetings in fifty-five languages, thirty-five sounds (natural and man-made), and portions of twenty-seven musical pieces, plus a picture of a breast-feeding mother. The message from Earth is recorded as analog wiggles in a spiral groove cut in a copper disk. As the record is spun, the grooves cause the needle to wiggle at audio frequencies reproducing the sounds with which it was made. Commercial records are made from vinyl plastic pressed into metal master molds. The *Voyager* record was made by the same methods used to make the metal masters, methods that did not fundamentally differ from those used in 1877 when Edison invented the phonograph. The messages also contained a two-square-centimeter area of ultrapure uranium 238, which has a half-life of 4.51 billion years and will allow any alien finding this "message in a bottle" to determine the elapsed time since launch.

So quickly is human technology advancing that were we to send this message today, we would use digital information instead of the analog method of an Edison phonograph record. It's hard to remember that desktop computers hardly existed when the *Voyagers* were launched. In another quarter of a century or less, it is reasonable that most human knowledge recorded in books and journals could be contained in digital storage weighing only a kilogram or so.

Yet Edison's technology has its advantages because it is less fragile than digital counterparts; thus, it will last longer under the abrasion of dust and gas. Astronomer Carl Sagan, who helped design these missives, estimated that the space-facing side of the metal phonograph record would be readable for a billion years, and the back side would last much longer.

We don't know if aliens could read our messages, but their mere presence might be as important as their contents. As Sagan wrote:

> But one thing would be clear about us: no one sends such a message on such a journey, to other worlds and beings, without a positive passion for the future. For all the possible vagaries of the message, they could be sure that we were a species endowed with hope and

perseverance, at least a little intelligence, substantial generosity and a palpable zest to make contact with the cosmos.

After *Voyager*, several other missions have also carried messages, though none are as elaborate. The *Stardust* mission that Don has helped design, which will collect comet dust and return it to Earth, will have a component that will stay in solar orbit and may eventually be hurled by passing planetary gravity outside the solar system. Its message—a silicon wafer that contains a few pictures and messages from the mission team, plus the names of 1.4 million people who signed up on the Internet—is embedded inside the spacecraft to shield it from dust impacts. Like the *Pioneer* pictures, it is a physical image that does not have to be decoded, but is so small that it has to be examined by a microscope. The letters of the names are only a micron (millionth of a meter) high and the width of the lines is only a tenth of a micron (or about one thousand atoms) wide. With this technology a hundred thousand books can be scribed onto a thin wafer of silicon the size of a piece of notebook paper. Any advanced civilization surely has technology such as electron microscopes, because microscopic imaging is an enabling technology that is necessary for development of microelectronics. As our own technology improves, our ability to send ever-longer messages in ever-smaller packages will undoubtedly advance.

Still, the chances anyone will find our bottles in the cosmic ocean are small. A more satisfying endeavor would be to send *us* in a bottle. Assuming that we cannot ever physically travel the galaxy in *Star Trek* fashion—we will discuss the difficulty of this in the next chapter—perhaps we can send some condensed form of us. In a sense, this has already been done. When the *Lunar Prospector* mission crash-landed on the Moon in 1999, it carried with it the ashes of planetary scientist Gene Shoemaker. He became the first-known organism to have permanently escaped the surly bonds of Earth. He is quite literally the man on the Moon.

In the future will we send away time capsules that contain not only our knowledge and history, but some of us too? Can we propagate our species beyond the life of our planet and Sun? A few grams of material could in principle carry DNA samples from every human on Earth. The sample could easily be kept cold and shielded for billions of years. Still, the difficulties are daunting. DNA is a complex molecule and its suitability for truly long-term storage is unknown. Even if we did send out a DNA capsule, it would never hit a planet just by chance. Space is so vast, and the emptiness between stars and planets so great, that an object traveling in space has almost no likelihood of impacting a planet within a trillion years. And after a trillion years, all of the stars of the Universe will have burned out and there will no longer be any habitable places left to inhabit! To have any chance of survival, DNA capsules would have to have a means to detect habitable places, maneuver to them, and then stop and land on them. Right now we lack such technology, but if it is ever developed perhaps the people of Earth might one day cast thousands or even millions of small DNA-bearing seedpods out into the galaxy.

Some scientists have speculated that such experiments elsewhere could have been the way life arrived on Earth.

In an even more speculative vein, perhaps our focus on DNA and biology is myopic and provincial and not really the way that life in the Universe works. Perhaps our future is totally different from the lives we enjoy in the twenty-first century. Perhaps all biological limitation will be overcome and we will become immortal, or even renewable, when tragic accidents happen. Perhaps we will become something analogous to computer chips. People as computer chips could easily be launched on paths that would endlessly journey among the stars. Some are convinced that biological or "wet life" is merely a beginning, and that life will evolve to much higher states. After all, people are already becoming used to replacing body parts with living and nonliving components. It could be only a matter of time until people become only brains married to machines. Some envision that human existence and even the soul will be uploaded into a computer's memory. The physical memory container could be exceedingly small and have very few environmental constraints. It (we) could last almost forever.

Physicist Frank Tipler has envisioned star-voyaging computers that bear the software encoding of thousands of real humans, accelerating to near light-speed in spaceships weighing a kilogram. If each spacecraft makes two copies of itself upon finding a new place to land, he estimates the entire Milky Way could be colonized in a million years, and the entire Universe in 10 million trillion years.

In explaining his scheme in a January 2000 article in *Wired* magazine, Tipler said that Einstein's laws of relativity require that spaceships be very small in order to make their acceleration to near light-speed practical, given that mass rises with velocity. Humans would become software, and "quantum computers could code an entire simulated city containing thousands of humans in only a few grams. Nanoscale (microscopic) von Neumann machines—capable of making any other machine—round out the payload. Powered by matter-antimatter annihilation, a 1-kilogram spaceship could reach Proxima Centauri, the nearest star, in only five years . . . human uploads have such a natural advantage over present-day people in the environment of space, it's exceedingly unlikely flesh-and-blood beings will ever engage in interstellar travel."

Intriguing. Still, however rational the prospect of seeding ourselves as DNA packets or software code, the romance and passion of life seems missing from such schemes. Is there a way we can escape the inexorable fate of planet Earth as flesh-and-blood human beings? Just as we dream of beating death with a fountain of youth or discovery of immortality, can we beat the grim life span of our home with science fiction schemes? After a look at the Drake Equation in the next chapter, we will look at some of the practical difficulties.

· 12 ·

# THE ENDS OF WORLDS
# AND THE
# DRAKE EQUATION

IN THE PREVIOUS CHAPTER WE LOOKED AT WHAT SORT OF legacy our *species* will leave. Yet each individual wants to believe that his or her time on Earth has some meaning, that each existence is not simply a life lived, ended, and forgotten. And if our life has any importance, then so too must our death, even if in the smallest way. Not only are there questions about the *meaning* of our lives—but surely there are uncertainties about our physical existence. After all, there are many potential fates for our bodies after we die. Most disintegrate away in the ground, leaving nothing behind. A few, through vagaries of preservation, have some parts that last for long periods of time, either as skeletons or even as mummies of some type. But these are rare cases. For most of us, within a very short period of time, there is no physical record that we existed at all.

What of planets, or even stars, for that matter? For our own star, the flaring into a red giant will be followed by a stellar retreat into a dwarf stage that will last untold billions of years. As astronomers gaze out into the heavens with their powerful telescopes, they see evidence that there are billions of such stellar tombstones. The galaxy is littered with dead stars, the markers of how many dead planets, and of how many dead civilizations that for a time circled these stars when they were young and vigorous?

Each of us humans is one of some 6 billion humans currently alive, one of the numerous humans that have existed on Earth since the formation of our species during the Pleistocene epoch. In an analogous fashion, our planet Earth is one of an even greater population, one planet among the billions in the galaxy, and untold billions and billions in the Universe. Does its existence—and ultimate passing—have any sort of meaning? For science it certainly does. Our new understanding about the life span—and ultimate death—of habitable planets, based on the history of our Earth, has important implications for understanding and estimating the frequency of life—and of intelligent species—in the cosmos.

The presence of these stellar graveyards are thought-provoking reminders that any estimate about the frequency of life in the Universe must take into account the fact that once evolved, life has a finite duration on any world. And, like the ultimate age of individuals, the life spans of life-covered planets in large part depends on a great number of factors.

We have only to look at one of our nearest planetary companions to see how those variables work. Of all planets beyond the Earth, Mars is by far the best known. It has been poked, prodded, examined, and measured by a variety of Earth- and space-borne instruments, including those many that have successfully and unsuccessfully either landed or crashed on its surface. A wealth of data now suggests that early in its history, while our Earth was still a chaotic and uninhabitable world, Mars may have been a benign world, of equable temperatures and almost planet-spanning oceans. It may as well have been a world with an atmosphere that included oxygen. All of these factors lead to an inescapable conclusion—that the early Martian conditions would have been favorable for the development of life. Some scientists have even suggested that life arose on Mars, and was then transported to Earth—indeed, that all life on our planet has a Martian origin, transported to Earth as microbial spores amid small meteors blasted off of the Martian surface and later impacting the Earth—just as the famous Allen Hills meteorite did. For several hundred million years or more these benign conditions may have lasted. In several hundred million years evolution can work wonders. Perhaps the first geologists sampling Martian sedimentary rocks older than 4 billion years in age will find not only the fossil remains of bacteria but the remains of more complex organisms as well. Perhaps the fossils of animals will be found. What would that scene be like—the swing of a rock hammer against a Martian outcrop, splitting a piece of ancient Martian shale—and the heart-stopping joy of finding a mollusk look-alike—or the bones of a fish equivalent? Yet even if life did attain such a rapid rise in complexity on Mars, it did not last, for Mars as an environment for life died quickly. Even as bacteria on Earth were readying for the rush to higher grades of life, Mars was dying or was already long dead—if life ever originated there at all. On Mars, the oceans seeped back into the planet or were lost to space, the oxygen in the atmosphere bound itself to rocks, and life died out.

Perhaps Mars and Earth are end members—one short, the other long-lived. Or perhaps the fates of planets in our solar system are not typical at all. Whichever, it is certain that all planets as abodes for life age through time, and as they change they eventually lose the ability to sustain life. Sometimes they do so over immense periods of time, sometimes it might be fast. Some die of old age, and some are killed off by cosmic catastrophe. But all end eventually. This is an unavoidable fact of any consideration of the frequency of life in the cosmos.

It has been said that the twentieth century produced two iconic mathematical expressions: Einstein's famous $e=mc^2$, and the Drake Equation. The latter is the expression created by astronomer Frank Drake in the 1950s to predict how many alien civilizations might exist in our galaxy. The point of the exercise was to estimate the likelihood of detecting radio signals sent from other technological civilizations.

The Drake Equation is simply a string of factors that, when multiplied together, give an estimate of the number of intelligent civilizations in the Milky Way galaxy. As originally postulated, the Drake Equation is as follows:

$$N^* \times f_s \times f_p \times n_e \times f_i \times f_c \times f_l = N$$

where:

$N^*$ = stars in the Milky Way galaxy
$f_s$ = fraction of Sun-like stars
$f_p$ = fraction of stars with planets
$n_e$ = number of planets in habitable zone
$f_i$ = fraction of habitable planets where life does arise
$f_c$ = fraction of planets inhabited by intelligent beings
$f_l$ = percentage of a lifetime of a planet that is marked by presence of communicative civilization

The validity or knowledge of the likely values for these various terms obviously varies enormously. When Drake first published the equation, most of the values were quite speculative. Then and now there exists a good estimate for the number of stars in our galaxy (between 200 million and 300 million). The number of star systems with planets, however, was very poorly known in Drake's time. Although many astronomers believed that planets were common, there was no theory that proved that star formation should include the creation of planets and, in fact, many believed that the formation of planetary systems was exceedingly rare. Yet beginning in the 1970s, scientists started to assume that planets are common—so common, in fact, that Carl Sagan estimated that an average of ten planets would be found around each star. Real evidence of the existence of any other planets did not emerge until the 1990s with the findings of extrasolar planets. Astronomers were surprised to discover that most solar systems do not have the larger planets far from the Sun in well-behaved, nearly circular orbits as ours does. Geoff Marcy, the world's leading planet finder along with colleague Paul Butler, noted: "For the first time, we have enough extra-solar planets out there to do some comparative study. We are realizing that most of the Jupiter-like objects far from their stars tool around in elliptical orbits, not circular orbits, which are the rule in our solar system."

Almost all of the Jupiter-sized objects found to date are located either in orbits much closer to their sun than Jupiter is to our Sun, or, if at a greater distance from their sun, with highly elliptical orbits (observed in nine of the seventeen so far detected). In such planetary systems, the possibility of Earth-like planets existing in stable orbits is low. When the "Jupiter" is close to its sun, it will have destroyed the inner rocky planets. With elliptical orbits the "Jupiters" will have caused disruption of planetary orbits sunward, causing smaller planets to either spiral into their suns or be ejected into the cold death of interstellar space.

To estimate the frequency of intelligent life, the Drake Equation hinges on the abundance of Earth-like planets around solar-like stars. Compared to giant planets it is difficult to observe smaller rocky planets orbiting around other stars. At this time we cannot ascertain whether such planets—which we believe are necessary for animal life—are common, rare, or somewhere in between. For this exercise, let us assume that rocky, Earth-like planets located within the geographic distance from a star resulting in temperatures allowing the presence of liquid water on the planetary surface—the so-called habitable zone—are common. How can our study of the various ends of the Earth be used to better understand the frequency of complex life in the Universe?

Many astrobiologists have concluded that the formation of microbial life on a planet within a habitable zone might be relatively easy—that, given the conditions of a liquid water–covered surface, many or most planets will yield life of some sort. If so, then millions to hundreds of millions of planets in the galaxy have potential for advanced life. However, getting from microbial life to advanced life requires time and a stability of conditions. Only under such circumstances of long-term planetary stability where large-scale temperature fluctuations do not take place can the jump to the more fragile forms of complex life take place. With these thoughts in mind, in our book *Rare Earth* we added some new terms to the Drake Equation:

$$N^* \times f_p \times n_e \times f_i \times f_c \times f_l = N$$

where:

$N^*$ = stars in Milky Way

$f_p$ = fraction of stars with planets

$n_e$ = number of planets in a star's habitable zone

$f_i$ = fraction of habitable planets where life does arise

$f_c$ = fraction of planets with life where complex metazoans arise

$f_l$ = percentage of a lifetime of a planet that complex metazoans are extant—what we might call the "habitable life span" of a planet

How can the habitable life span of a planet be estimated? To answer such a question we first need a sense of what it is that a planet needs in order to evolve complex life in the first place. Perhaps complex life can occur on any planet where life evolves (or upon which life arrives) that also bears freestanding liquid water or some equivalent solvent. The example of Mars, given above, could illustrate such a case. But our Mars example also suggests that the life span of the biosphere on any such smaller planet would be short.

One of the Earth's most basic life-supporting attributes is its location, its seemingly ideal distance from the Sun. In any planetary system there are regions—distances from the central star—where the Earth could survive with a surface environment similar to its present state. This is the habitable zone, the region in a planetary system where habitable Earth clones might exist. Since its first introduction, the concept of the habitable zone has been widely adopted and has been the subject of several major scientific conferences, including one held by Carl Sagan in one of the final scientific contributions of his brilliant career, just before his untimely death. An important aspect of the habitable zone is the subject of this book: planets, such as our Earth, age and ultimately "die" as abodes for life. Many issues are involved but one is that the zone of habitability moves outward with time, leaving planets behind. We cannot exclude this factor from our estimate of the frequency of life in the cosmos.

Our closest neighbors in space provide sobering examples of what happens to planets inside of, or outside of, the habitable zone. Interior to the HZ a planet gets too hot. Venus is an example where this has happened. The surface of Venus is nearly hot enough to glow. If it ever had an ocean, it has evaporated and been totally lost to space.

Outside of the habitable zone, temperatures are too low. Mars is outside of the HZ, and is frozen to depths of many kilometers below its surface. If the Earth was moved outward (or if the Sun reduced its energy output), its atmosphere would cool to a point where it would become ice covered. We have elsewhere proposed that entire galaxies, as well as individual stellar systems, have regions both more or less suited for the long-term maintenance of life. We named these geographic regions galactic habitable zones. Unlike a star system, where habitable planets exist in regions dictated by temperature, the habitable zones of any galaxy are dictated by energy—which in too large a dose creates mass extinction—and metal, which in too low a concentration precludes the survivability of life on any planet.

Our galaxy is a spiral galaxy, one of the three major galaxy types (the others being elliptical and irregular galaxies). In any galaxy the concentration of stars is highest in the galactic center, and then diminishes outward away from the center. Spiral galaxies are dish-shaped, with branching arms when viewed from the top. But if viewed from the side they are quite flat, with a thickness of barely 1 percent of the overall diameter. Our galaxy has an estimated diameter of about one hundred thousand light-years. Our Sun is about twenty-five thousand light-years from the center, in a region between spiral arms where star density is quite low compared to the more crowded interior. In this position we slowly orbit around the central axis of the galaxy. Like a planet revolving around a star, we maintain roughly the same distance from the galactic center—which is a good thing. Our star—through great good luck—is located in the habitable zone of the galaxy. The inner margins of this galactic habitable zone are defined by the density of stars (and the dangerous supernovae) and the energy sources found in the central region of our galaxy, while the outer regions of habitability are defined by something quite different: not the flux of energy, but the type of matter found. The habitable zone is perhaps a narrow region in any galaxy.

At the present time we can estimate only roughly the limits of this habitable zone region. Its inner region is surely defined by celestial catastrophes occurring closer to the center, but we cannot yet estimate how close to the center of the galaxy that line is. Perhaps it extends ten thousand light-years from the center, perhaps more. However, we do have at least a vague idea of the forces producing this inner limit. Life is a very complex and delicate chemical balancing act, easily destroyed by too much heat or cold—and too many gamma rays, X rays, or other types of ionizing radiation. The center of any galaxy produces all of these.

Among the most numerous of the lethal stellar members of any galaxy are a type of neutron star called magnetars. These are collapsed stars of small size but astonishingly high density that emit X rays, gamma rays, and other charged particles into space. Because energy dissipates as a square of distance, these objects are of no threat to our planet because of their great distance from us. Closer to the center of the galaxy, however, their frequency increases. Any galactic center is a mass of stars, some of which are lethal neutron stars, and it seems unlikely that any form of life as we know it could commonly exist.

An even greater threat comes from exploding stars, the super-novae discussed in a previous chapter. Any star going supernova would likely sterilize life within a one light-year radius of the explosion, and affect life on planets as far as thirty light-years away. The very number of stars in galactic centers increases the chances of a nearby supernova. Our Sun and planet are protected simply by the scarcity of stars around us.

The final risk is that of asteroid or comet impact. In the more central, and crowded, regions of the interiors of galaxies, neighboring stars can be close enough to perturb comets and cause them to fall inward into the interior of the stellar system. Such comets have planet-crossing paths, and are thus on courses that can cause them to impact planets revolving around the central star. Such impacts, based on the terrestrial record, are inimical to life and can bring about planetary sterilization if the impact is sufficiently energetic, which is a function of impactor size and velocity.

The outer limit of the galactic habitable zone is defined by the elemental composition of the galaxy. In the outermost reaches of the galaxy the concentration of heavy elements lowers because the rate of star formation and thus element formation is lower. There is a continued decrease in the relative abundance of elements heavier than helium as one goes outward from the centers of galaxies. The abundance of heavy elements is probably too low to form terriginous planets as large as the Earth. Important attributes of our planet are its size, its radioactive heat, and the presence of a large metal core. All appear necessary to produce animal life: the metal core produces a magnetic field that protects the surface of the planet from radiation from space, while the radioactive material maintains the engine of plate tectonics, also, in our view, necessary for maintaining animal life on the planet. Planets such as Earth are probably rare in the outer regions of the galaxy.

The ideas presented above are in their infancy. Yet even in this early stage we are beginning to understand those factors necessary for the longevity of habitable planets. In order to understand such longevity, we need to know the following factors affecting habitability through time:

Solar luminosity through time (and mass of star)

Mean global planet temperature

Presence or absence of plate tectonics

Presence or absence of a rock-silicate feedback cycle

Continental area

Biological productivity

Surface heat flow

Geographic position of the planetary system in its galaxy

The numerous ends of the Earth and its varied systems of life that we have detailed in this book are thus not necessarily specific to our planet. They can be used to better understand the fates of our own world and of habitable worlds in general. We would like to know what the life span of animal life—what we call metazoans—will be on a variety of worlds. Where stars and planets are smaller, or larger, or located more centrally or more marginally in the galaxy—what will the fate of planets be? How long can they remain habitable? This is why we study planetary health—including that of our own planet Earth.

# THE GREAT ESCAPE

HUMANS HAVE A HABIT OF MOVING ON WHEN THINGS GET bleak. The frontier has been a place of new beginnings. So it seems only natural that we will ultimately fly to other worlds. After all, ancient Polynesians traveled thousands of miles in open canoes across totally unknown seas to discover Hawaii. Surely our descendants, with all of their knowledge and technology, can travel to distant planets. Why not? The very reason the lands of Earth are so well populated is that Earth's creatures naturally strive to expand their territories. Won't our descendants do the same, in deep space?

In only a century we advanced from the horse and buggy era to a world of airplanes, miracle drugs, computers, lasers, hydrogen bombs, the Internet, and rockets that can fly to distant planets. In coming centuries won't technology take us to other planets, to other stars, and even, with full warp speed, to other galaxies?

Not necessarily. Technological progress is not open ended. There are reasons to be pessimistic about future advances as well as optimistic. The laws of physics and chemistry are universal, and they do not adjust themselves to meet our whims or needs. Moore's rule that computer speed doubles every eighteen months cannot be extrapolated forever into the future: the speed of light and the limits of miniaturization to atomic scale put a limit to how fast and small circuitry can go. We will never build bicycles that can be ridden much faster that they can be at present, and we will never see an airplane that will fly in the lower atmosphere much faster than the SR-71 spy plane, designed in the 1950s. There will certainly be fabulous and unforeseen advances in the future, but there are real physical limits on what can be done. Truly miraculous developments like walking on water or traveling beyond the speed of light would require violating natural laws that we know to exist. Talk of using "wormholes" to tunnel to the other side of the Universe in instantaneous fashion is, at this point, just that: talk. Millions of years into the future, the work of Newton and Einstein will still be studied in physics classes, and force will still equal mass times acceleration. The basic laws are universal, they are the way things are, and they are not products of the human mind. Because of this, escape may not be as easy as portrayed in science fiction.

IF WE ARE ABLE TO GO ELSEWHERE, WHERE WOULD WE GO? A common answer is Mars, the most Earth-like planet in the solar system and a close neighbor. It is 50 percent farther from the sun than the Earth is, and the sunlight there is less than half as bright. When the Earth gets too hot why not go to Mars, surely a cooler place? Many people believe that this is inevitable. Surely Earth will have already colonized Mars! The final move will just be like packing the bags for a summer cabin in the boondocks. While Mars is a small planet, less than half the diameter of our own, its surface area is comparable to the total land area of Earth. Ideally it should be able to house Earth's hordes.

Well, maybe. Although a wonderful planet with fascinating geography, Mars is a most inhospitable place for humans. The surface pressure is similar to the air on Earth at 30 kilometers, or more than three times the height of Mount Everest. There is no oxygen. There is no food. There is no surface water, and it never gets warm. Deadly ultraviolet light from the Sun bakes the land.

Still, we have gone to the Moon, an even more hostile environment. Surely Mars is the next logical place to go on our road to the stars. We permanently inhabit the South Pole and an orbiting space station. Why not the red planet? If astronauts go there and come back, will this lead to the establishment of a station, a colony, and a future home for displaced Earthlings?

The problem is not technology per say, it is the cost. Apollo went to the Moon in 1969 and, over the course of less than three years, six missions and twelve Earthlings landed on the lunar surface. The cost for the program, in equivalent present dollars, was more than $100 billion. The country that funded Apollo was the most prosperous in Earth history, and although it could go back it has not been motivated to do so since 1972.

Going to Mars is much more difficult and dangerous than going to the Moon. You can get to the Moon in three days, but it takes months to reach Mars. Round-trip mission times are on the order of a year or more. Astronauts are exposed to potentially lethal cosmic rays from the Sun and have to survive long periods of low gravity, an environment that produces irreversible loss of bone mass. The degradation of the human body in space has been feverishly worked on for more than forty years, and yet there is still no clear solution to the problem of long-term survival of humans in space.

The hope is that Mars can be terraformed to make it more Earth-like. The general idea is to genetically engineer organisms that could live on Mars and, over time, produce an atmosphere that would be thick enough to allow liquid water to remain on the surface. Ultimately oxygen could be produced and Mars might become a respectable Earth clone, or at lease a place where people could live for long periods of time. Will this ever happen? It is possible, but there are reasons to suspect it will not be practical. Earthlings may not want, or be able, to pay the cost. We have difficulty

sustaining the environment of our home planet, let alone successfully manipulating that of another. Would anyone really want to
relocate to a planet with half our gravity and a Sun half as bright?
Even on our own crowded planet, vast reaches of the Arctic and
Sahara remain almost uninhabited. Will humans persist in the hundreds or thousands of years it would take to terraform Mars?

Not only is planet modification a monumental task, but spacecraft operations in deep space are not nearly as easy as shown in the
movies. They are highly complex machines prone to fail far from
home. Airplanes have been made safe and even economical, but the
enormous number of technological achievements that made this
possible were obtained by phenomenal expenditures of money,
resources, and human life. Nearly all the technology that enables
air travel was developed by the nearly unlimited largess of military
budgets in a world full of countries desperately preparing for war.
Space technology (along with electronics, computers, the Internet,
and many other things that we take for granted) was also developed by defense budgets.

Even if we had a war to spur development, space travel will never
be as easy and successful as air travel. Spacecraft are not mass-
produced, and even though, like airplanes, they include redundant
systems and fault-tolerant designs, they cannot afford to fail. They
cannot turn around and land at the nearest airfield or plowed field.
All lethal problems to spacecraft must be thought out in advance.
This is impossible to achieve. Consider the *Hubble Space Telescope*. Even with the greatest of care and a budget of billions of
dollars, *Hubble* was put into space with a misshaped mirror, solar
panels that flapped every time they passed in and out of the sunlight, and gyroscopes that failed at a much higher rate than
expected. *Hubble* became a huge success only because it was so
close to the ground that the space shuttle could carry up heroic
astronaut crews on repeated repair and refurbish missions. In deep
space, repair missions are not practical.

Mars, in fact, has been an excellent testing ground for the reliability of complex space missions. Since 1959, thirty-five missions have been sent to Mars. Over half have failed. As technology has continued to advance, the success rate has not improved. Can we ever make traveling to Mars as safe as traveling on a plane or even flying in the space shuttle? Probably not.

An additional obstacle to Martian exploration and ultimate habitation is power. Humans need lots of power to keep warm, eat, take baths, breathe air, work, and occupy their time. Power on Earth is largely derived from coal, natural gas, and oil. The only known sources of power on Mars are sunlight and wind. Perhaps future explorers can make huge solar energy collectors on Mars, but they may not be adequate to supply a colony with mining and water-collection needs. Nuclear power may be the only alternative, but nuclear technology is currently almost dormant. Will people of the future have the knowledge, the materials, or even the permission to develop nuclear power for Mars?

As long as the human population remains prosperous, motivated, and technologically capable, it is likely that people will go to Mars. It remains a huge question, however, whether it will ever be possible for large numbers of people to be transported there or live there.

The real irony is that it doesn't really matter in the long run. The same Sun that will doom life on Earth may eventually fry Mars as well. Our neighbor simply isn't far enough away to escape the bleak future portrayed in this book. Travel to Mars can prolong the survival of *Homo sapiens,* but the end will still be in sight.

So, if Mars is not a permanent solution, are there other planets or moons farther from the Sun that could harbor human life? The prospects do not look good. There are several places that might support microbial life, but hosting millions of humans is an entirely different matter. Europa, the second large moon out from Jupiter, is often mentioned as a likely place for life. There is very strong evidence that Europa has an ocean of water beneath its fantastic surface of grooved and fractured ice. Yet even if life in the ocean is possible, it is not so on Europa's surface. The moon has only a wisp of an atmosphere and the deadly radiation from the radiation belts around Jupiter produce a surface environment similar to that inside a nuclear reactor.

Other famous places that might possibly support life also seem to be unlikely sites for future human colonization. Titan, Ganymede, Callisto, and Triton are all quite different bodies from Earth. It is more likely that humans will transform into some strange kind of superbeings that could thrive in hostile environments than it is that these distant moons could be made at all Earth-like.

Perhaps the solution is not to try to replicate the Earth on a planet or moon, but to move the Earth. Moving our planet away from the Sun is a daunting task, but probably much easier than moving people to other planetary systems. We have much of the necessary technology needed to move the Earth. In a 2001 research paper by Don Korycansky, Gregory Laughlin, and Fred Adams, it was suggested our world could use multiple near-misses by a comet 100 kilometers in diameter to nudge the Earth outward at just the right rate to compensate for the increase in brightness in the Sun. Their clever scheme uses gravity assist, a method already harnessed by both spacecraft and comets. A body flying past a planet gravitationally interacts with the planet and either adds or subtracts from its orbital energy. With the right encounter geometry, energy is added and the planet moves slightly outward. The proposed scheme has a single comet repeatedly flying past Earth, each time moving Earth a little farther from the Sun. The comet then loops far past the orbit of Pluto and comes back, passing Jupiter and Saturn along the way. The comet loses energy during its Earth encounters, but gets it back from its Jupiter encounters. Jupiter would be nudged inward, but by a much smaller amount because it is so much more massive than our planet. In theory, Earth could be scooted just enough farther from the Sun so that habitable surface temperatures could be maintained.

During the next 6.3 billion years the Sun will more than double in brightness, but this can be compensated for by moving Earth about 50 percent farther from the Sun than it is now, or to approximately the orbital distance of Mars: a bit of astronomical engineering effort that could extend the age of animals to 6 billion years from now. Beyond that time, of course, the Sun becomes a red giant and the Earth would get fried at even Martian distances: but this idea could buy life on our planet billions of years of extra time.

The comet-Earth flybys would occur about every six thousand years, and a million of them would be required to move Earth all the way to Mars. (Or, rather, where Mars was. Mars would have to go because there is not room for two planets in the same orbit. Another minor problem.) The operation of a 100-kilometer-diameter comet as a spacecraft requires some energy to provide small velocity changes to make sure that the precise orbital path is maintained, but the amount of energy needed is small and could be supplied by a nuclear reactor using deuterium and tritium (the stable and radioactive heavy isotopes of hydrogen) as fuel. Each Earth flyby would require mining a 2-kilometer cube of ice from the comet, extracting the heavy hydrogen and converting it to energy to provide rocket propulsion. All of this work would be done far beyond the orbit of Pluto.

Can this be done? Yes, in theory. In practice, it may be even more difficult than terraforming Mars.

The comet is large and zooms past Earth at 40 kilometers a second. If something goes wrong and it hits Earth, it carries enough energy to kill all life on the planet. The encounters are done not just once, but a million times. Can any human activity be perfectly maintained for billions of years? Who runs the show, who assures that everything is done perfectly, who pays the bills? There are two highly pertinent quotes from the paper:

Our initial analysis shows that the general problem of long-term planetary engineering is almost alarmingly feasible using technologies that are currently under serious discussion. The eventual implementation of such a program would profoundly extend the time over which our biosphere remains viable.

An obvious drawback to the proposed scheme is that it is extremely risky and hence sufficient safeguards must be implemented. The collision of a 100 km diameter object with Earth at cosmic velocity would sterilize the biosphere most effectively, at least to the level of bacteria. This danger cannot be overemphasized.

The risk is not limited to the impact of a large comet. The repeated encounters are likely to cause Earth to loose its Moon. This would not only be a shame, but a great danger. With the Moon gone Earth's spin axis would begin to wander and drive unknown climate changes. Moreover, while the Moon would be pushed away from its Earth-circling orbit, it would orbit the Sun in an Earth-crossing orbit. Left on its own, the Moon would be likely to smack into the Earth, causing far more damage than the impact of a measly 100-kilometer comet. There is also the issue of moving Mars to make room for Earth in its orbit. If the move is messed up and Mars impacts Earth, the consequences would be vastly more severe than even the lunar impact.

This is heady stuff. Great risk but great gain. Desperate life on a dying planet would surely try this type of thing if it could. But is it really possible? Only time, long time, will tell.

MOVING EARTH SEEMS LIKE A LONG SHOT, BUT SUCH AN endeavor may be trivial compared to that staple of *Star Wars* and *Star Trek:* interstellar travel. Many people feel it is our destiny to travel to the stars. Thirty-five percent of the American public believes that alien creatures similar to humans have visited Earth or are here at the present time. Yet traveling to the stars is an extraordinarily difficult thing to do. "Practical" travel to the stars is totally beyond the ability of known technology.

Any dot.com millionaire with $50 million or so to spare can travel to the nearest stars, but the trip would be neither practical nor pleasant. For this level of money you can buy a Delta rocket and have yourself launched on a trajectory that would fly you by Jupiter, where its gravity would bend your orbit onto an escape path from the solar system. Unfortunately, the trip to the nearest star would take more than ten thousand years and there would be no bathroom, no food, and no water. You would have no real means of making midcourse corrections to ensure that you passed close to the target star. Once there, you would not have the means to stop and so would fly by, probably not coming close to another star for a hundred million years. Even if you *could* stop, and the star had planets, you very likely wouldn't like their environments: it would be highly unlikely that any of them would closely resemble Earth. But if you didn't like the destination, there would be no refund and no way back.

The difficulty of "practical" travel between stars is getting there on the timescale of the human life span. It's easy if you're not in a hurry, but if you want to get there quickly and do fancy things like stop, the task is nearly impossible. Serious attempts to design spaceships to travel to the stars have not produced encouraging results. Even if we could make rockets propelled with antimatter fuel, and solve all the accompanying nasty details, it would still take all of the Earth's energy output for centuries just to make the fuel. Beaming energy with lasers, interstellar ramjets, and many other great suggestions are interesting to think about, but unlikely. It is probably much easier to travel to the center of the Earth than to voyage to the nearest star. The starships of TV, movies, and novels are products of wishful thinking. Interstellar travel will likely never happen, meaning we are stranded in this solar system forever. We are also likely to be permanently stuck on Earth. It is our oasis in space, and the present is our very special place in time. Humans should enjoy and cherish their day in the Sun on a very special planet and not dwell too seriously on thoughts of unicorns, minotaurs, mermaids, and the *Starship Enterprise*.

Our experience on Earth is probably repeated endlessly in the cosmos. Life develops on planets but it is ultimately destroyed by the light of a slowly brightening star. It is a cruel fact of nature that life-giving stars always go bad.

# Epilogue

HUMAN BEINGS ARE BORN, THEY GROW, THEY LIVE, AND THEY
die. In this book we have endeavored to show that habitable planets like our own have that same inevitable, joyous, tragic, and fulfilling life span. If a planet is close enough to its star for life to arise, it is close enough for that life eventually to be destroyed by that star. Moreover, it will be extremely difficult for the planet's inhabitants to escape to another world.

Obviously, this is as bleak a prognosis as that of our own individual mortality. In our book *Rare Earth*, we endeavored to show that Earth-like planets and advanced forms of life are likely to be uncommon in our galaxy and Universe, however much we wish it were not so. The conditions that sustain life here represent an extraordinary balance, a combination of happy circumstances that is rare indeed. In this book we have attempted to show that not only is our planet wonderfully unusual, but that its nourishing environment has a time limit. The world will come to an end, scientifically, and in several sequential ways. We live in a glorious summer of beauty, diversity, and resources. It will not always be so.

Prophecy is a risky business, of course. We've made things a little easier on ourselves by forecasting thousands, millions, and even billions of years into the future, not tomorrow when we—like the hapless meteorologist—might be proven wrong. Still, this book is not idle speculation. Our thesis is that the recent discoveries of how the Earth worked in the past and how it works today can clearly suggest its fate in the future. We believe that over time, our planet will revert to its original, more primitive conditions. Its history is a parabola, and we are enjoying Earth somewhere near— perhaps a bit after—its biological peak. If nothing else, we should feel very, very lucky.

Just how to react to our prognosis is more difficult than the prognosis itself. If first civilization, then life, and finally the planet itself are doomed to destruction, what does that mean? For that matter, the prevailing view in cosmology today is that the Universe will continue to expand and run down indefinitely until it is filled with nothing but dead stars and lifeless planets and, finally, trillions of years into the future, disassociated atoms. While some scientists theorize our Universe will eventually recollapse and be reborn in a second Big Bang, evidence to date suggests a much more prosaic and desultory fate, like the Energizer Bunny finally running out of batteries and coming to a lonely, pointless halt. The preponderance of astronomical evidence to this point is that, quite literally, the Universe will run out of gas and go dark.

This expectation is more than a little disquieting: for philosophy, for religion, and for hope.

As scientists, we have no duty to give meaning to such a forecast: we just tell it like it is, to the best of our knowledge. Nor do we have special expertise for such meaning. This book is a summary of our best model of our planet's fate based on recent research, and driven by our own natural curiosity. We leave the philosophic, religious, and moral implications to you.

Still, we're also human beings who entered science with a very human sense of wonder about the extraordinary world and Universe we live in. We realize that scientific discovery (witness Darwin, Einstein, or Oppenheimer) has a profound impact on human society and thought. It's hard to decide which is more daunting for the human mind to grasp: that this green world will end, or that it will take immense stretches of time—time virtually unfathomable to the puny minds of our brief existences—to reach that end. This planet is not immortal, but neither is it likely to end anytime soon, much to the frustration of apocalyptic preachers. There would seem to be a lesson here, but what is it?

One truth is that this moment on this Earth truly is a precious gift, to be savored and appreciated. If we heedlessly destroy this world, it is unlikely we will find another to replace it. Or be able to get to any refuge, even if we could find it.

Another obvious lesson is that we tinker with our atmosphere and oceans at grave risk. Yes, Earth has developed an astonishingly robust and reinforcing cycle of rock, water, air, and life to keep itself a biological haven. Yes, its environment is to a certain degree resilient, and strives to keep temperature in balance. But we've also shown in this book how those systems have broken down in the past and will likely do so in the future, with mass extinctions a possible result. We are still near the beginning of understanding how Earth systems work, and what must be done to sustain their health. What we have learned so far is that it is foolish to take those systems for granted. Right now we are pumping carbon dioxide into the atmosphere at truly heedless rates, despite overwhelming evidence that $CO_2$ concentrations have been a critical factor in tipping our planet's climate periodically toward disaster. Political leadership on this issue has been disappointingly slow.

It is possible that this book will encourage additional research, discussion, and speculation about our species' long-term future. We've been dismissive in these pages of belief in easy escapes from Earth's life cycle because we believe fantasies of such escape encourage indifference to immediate and future perils. "Earth? Who needs it? We'll just move on." But we are not dismissive at all of thinking about escapes, or solutions, or alternatives. The past hundred years or so have seen an explosive expansion of human perspective, from our growing awe at the dimensions of the cosmos to the baffling paradoxes of the subatomic world. Does our species have a long future? Where? How? We want this book to be one starting point for those questions, not an end point.

The careful reader will have noticed that we have refrained in these pages from predicting precisely when human beings leave the story of our planet. There are two reasons for this. One, of course, is that we don't know. And secondly, such a question is not for us to answer. It is really for you readers collectively to answer, and act upon, as you improve your understanding of the life and death of planet Earth. It is our hunch that humans, though limited in number, may be among the last animals to reach extinction.

IT IS THE YEAR 7 BILLION A.D. THE SUN HAS GONE INTO ITS RED giant phase. The Earth has been consumed by the outer envelope of the 100-million-mile-diameter sun. Mars is a dried and lifeless body with a surface temperature sufficient to melt its crustal rocks. Jupiter is a roiling, heated mass rapidly losing gas and material to space. The ice cover of Jupiter's moon Europa has long since melted away, followed by the disappearance of its oceans to space. Farther away, Saturn has lost its icy rings. But one world of this vast solar system has benefited from the gigantic red orb that is the Sun. It is Saturn's largest moon, aptly named Titan.

Long before, in the time of humanity, a science fiction writer named Arthur C. Clarke penned a series of tales about the moon of Jupiter named Europa. In these stories, alien beings somehow turned Jupiter into a small but blazing star, and in so doing warmed Europa—and brought about the creation of life. A wonderful, though physically impossible, fable. Now, in these late days of the solar system, the huge red Sun was doing the same to Titan, changing it from frozen to thawed, and in so doing liberating the stuff of life. But Titan was always a very different world than Europa. Like Europa, Titan always had oceans, frozen, to be sure, but oceans nevertheless. But where Europan oceans were water, those of Titan were of a vastly different substance—ethane. Titan had always been covered with a rich but cold stew of organic materials. And with the coming of heat, for the first time Eden came to Titan. Like a baby born to an impossibly old woman, life came to this far outpost, the last life ever to be evolved in the solar system.

The red giant phase was short-lived—only several hundred million years, in fact. But it was enough. For a short time, for the last time, life bloomed in the solar system. After death, once more came the resurrection of life in masses of tiny bacteria like bodies on a moon once far from a habitable planet called Earth, a place that, in its late age, evolved a species with enough intelligence to predict the future, and be able to prophesize how the world would end. . . .

# · S O U R C E S ·

Abe, Y. and T. Matsui. 1988. "Evolution of an Impact-Generated $H_2O$-$CO_2$ Atmosphere and Formation of a Hot Proto-Ocean on Earth." *J. Atmos. Sci.* 45: 3081–310.

Achenbach, J. 1999. *Capured by Aliens: The Search for Life and Truth in a Very Large Universe.* New York: Simon & Schuster.

Adams, F. and G. Laughlin. 1999. *The Five Ages of the Universe: Inside the Physics of Eternity.* New York: Free Press.

Alley, R. B. and M. L. Bender. 1998. "Greenland Ice Cores: Frozen in Time." *Scientific American* 278: 80–85.

Armstrong, R. L. 1991. "The Persistent Myth of Crustal Growth." *Aust. J. Earth Sci.* 38: 613–30.

Arrhenius, G. 1985. "Constraints on Early Atmosphere from Planetary Accretion Processes." *Lunar and Planetary Sciences Institute Rep.* 85-01: 4–7.

Atwater, T. 1970. "Implications of Plate Tectonics for the Cenozoic Tectonic Evolution of Western North America." *Geol. Soc. Amer. Bull.* 81: 3513–36.

Benarde, Melvin A. 1992. *Global Warning–Global Warming.* New York: Wiley.

Berner, R. A., A. C. Lasaga, and R. M. Garrels. 1983. "The Carbonate-Silicate Geochemical Cycle and Its Effect on Atmospheric Carbon Dioxide over the Past 100 Million Years." *Am. J. Sci.* 283: 641–83.

Berner, R. A. and A. C. Lasaga. 1989. "Modeling the Geochemical Carbon Cycle." *Scientific American* 260(3): 74.

Berner, R. A., 1991. "A Model for Atmospheric $CO_2$ over Phanerozoic Time." *Am. J. Sci.* 291: 339–76.

Berner, R. A. and D. M. Rye. 1992. "Calculation of the Phanerozoic Strontium Isotope Record of the Ocean from a Carbon Cycle Model." *Am. J. Sci.* 292: 136–48.

Berner, R. A. 1992. "Weathering, Plants, and Long-term Carbon Cycle." *Geochim. Cosmochim.* 56: 3225–31.

Berner, R. A. 1993. "Paleozoic Atmospheric $CO_2$: Importance of Solar Radiation and Plant Evolution." *Science* 261: 68–70.

Berner, R. A. 1994. "Geocarb-II—a Revised Model of Atmospheric $CO_2$ over Phanerozoic Time." *American Journal of Science* 294: 56–91.

Berner, R. A. 1997. "The Rise of Plants and Their Effect on Weathering and Atmospheric $CO_2$." *Science* 276: 544–46.

Bilger, Burkhard. 1992. *Global Warming.* New York: Chelsea House Publishers.

Bolin, B. 1986. *The Greenhouse Effect, Climatic Change, and Ecosystems.* New York: Wiley.

Bounama, C., S. Franck, and W. von Bloh. 2001. "The Fate of Earth's Ocean." *Hydrol. Earth Syst. Sci.* 5(4):569–75.

Bowring, S. A. et al. 1993. "Calibrating Rates of Early Cambrian Evolution." *Science* 261: 1293–98.

Broecker, W. S. 1983. "The Ocean." *Scientific American,* 249: 79–89.

Broecker, W. S. 1994. "Is Earth Climate Poised to Jump Again?" *Geotimes,* Nov. 1994, pp. 16–18.

Broecker, W. S. 1999. "What If the Conveyor Were to Shut Down? Reflections on a Possible Outcome of the Great Global Experiment." *GSA Today* 9(1):1–7.

Caldeira, K. and J. F. Kasting. 1992. "The Life Span of the Biosphere Revisited." *Nature* 360: 721–23.

Carroll, S. B. 1995. "Homeotic Genes and the Evolution of Arthropods and Chordates." *Nature* 376: 479–85.

Cerling, T. E., J. M. Harris, B. J. MacFadden, M. G. Leakey, J. Quade, V. Eisenmann, and J. R. Ehleringer. 1997. "Global Vegetation Change Through the Miocene-Pliocene Boundary." *Nature* 389: 153–58.

Cerling, T. E., J. R. Ehleringer, and J. M. Harris. 1998. "Carbon Dioxide Starvation, the Development of C4 Ecosystems, and Mammalian Evolution." *Phil. Trans. R. Soc. Lond.* 353: 159–71.

Chapin, F. Stuart. 1992. *Arctic Ecosystems in a Changing Climate: An Ecophysiological Perspective.* San Diego: Academic Press.

Christensen, U. R. 1985. "Thermal Evolution Models for the Earth." *J. Geophys. Res.* 90: 2995–3007.

Clarke, A. C. 1985. *Profiles of the Future.* New York: Warner Books.

Clarkson, Judith and Jurgen Schmandt. 1992. *The Regions and Global Warming: Impacts and Response Strategies.* New York: Oxford University Press.

Cochran, M. F. and R. A. Berner. 1992. "The Quantitative Role of Plants in Weathering" in *Water-Rock Interaction,* eds. Y. K. Kharaka and A. S. Maest. Aldershot, England: Ashgate Publishing, pp. 473–76.

Condie, K. C. 1984. *Plate Tectonics and Crustal Evolution,* 2nd ed. Oxford: Pergamon Press.

Cotton, William R. and Roger A. Pielke. 1995. *Human Impacts on Weather and Climate.* New York: Cambridge University Press.

Covey, C. 1984. "The Earth's Orbit and the Ice Ages." *Scientific American* 250: 58–66.

Coward, Harold G. and Thomas Hurka. 1993. *The Greenhouse Effect: Ethics and Climate Change.* Waterloo: Wilfrid Laurier University Press.

Cox, A. 1973. *Plate Tectonics and Geomagnetic Reversals.* San Francisco: W. H. Freeman and Co.

Crimes, T. P. 1994. "The Period of Early Evolutionary Failure and the Dawn of Evolutionary Success: The Record of Biotic Changes Across the Precambrian-Cambrian Boundary," in *The Paleobiology of Trace Fossils,* ed. S. K. Donovan. London: Wiley, pp. 105–33.

Crutzen, Paul J. and T. E. Graedel. 1995. *Atmosphere, Climate, and Change.* New York: Scientific American Library.

Dalziel, I. W. D. 1992. "On the Organization of American Plates in the Neoproterozoic and the Breakout of Laurentia." *GSA Today* 2: 237.

DePaolo, D. J. 1984. "The Mean Life of Continents: Estimates of Continental Recycling from Nd and Hf Isotopic Data and Implications for Mantle Structure." *Geophys. Res. Lett.* 10: 705–8.

Dietz, R. S. 1961. "Continent and Ocean Basin Evolution by Spreading of the Sea Floor." *Nature* 190: 854–57.

Dornbusch, R. and J. Poterba. 1991. *Global Warming.* Cambridge, Mass: MIT Press.

Ehleringer, J. R., and T. E. Cerling. 1995. "Atmospheric $CO_2$ and the Ratio of Intercellular to Ambient $CO_2$ Levels in Plants." *Tree Physiol.* 15: 105–11.

Ehleringer, J. R., T. E. Cerling, and B. R. Helliker. 1997. "C4 photosynthesis, Atmospheric $CO_2$, and Climate." *Oecologia* 112: 285–99.

Ephraums, J. J., J. T. Houghton, and G. J. Jenkins. 1990. *Climate Change: The IPCC Scientific Assessment.* New York: Cambridge University Press.

Erwin, D. H. 1993. "The Origin of Metazoan Development." *Biological Journal of the Linnean Society* 50: 255–74.

Eyles, N. 1996. "Passive Margin Uplift Around the North Atlantic Region and Its Role in Northern Hemisphere Late Cenozoic Glaciation." *Geology* 24: 103–6.

Firor, John. 1990. *The Changing Atmosphere: A Global Challenge.* New Haven: Yale University Press.

Fisher, David E. 1990. *Fire and Ice: The Greenhouse Effect, Ozone Depletion, and Nuclear Winter.* New York: Harper & Row.

Flavin, C. 1989. *Slowing Global Warming: A Worldwide Strategy.* Washington, D.C.: Worldwatch Institute.

Franck, S. and I. Orgzall. 1988. "High-pressure Melting of Silicates and Planetary Evolution of Earth and Mars." *Gerlands Beitr. Geophys.* 97: 119–33.

Franck, S. 1992. "Olivine Flotation and Crystallization of a Global Magma Ocean." *Phys. Earth Planet. Inter.* 74: 23–28.

Franck, S. and C. Bounama. 1995. "Effects of Water-Dependent Creep Rate on the Volatile Exchange Between Mantle and Surface Reservoirs." *Phys. Earth Planet. Inter.* 92: 57–65.

Franck, S. and C. Bounama. 1995. "Rheology and Volatile Exchange in the Framework of Planetary Evolution." *Adv. Space. Res.* 10: 79–86.

Franck, S. and C. Bounama. 1997. "Continental Growth and Volatile Exchange During Earth's Evolution." *Phys. Earth Planet. Inter.* 100: 189–96.

Franck, S., K. Kossacki, and C. Bounama. 1999. "Modelling the Global Carbon Cycle for the Past and Future Evolution of the Earth System." *Chemical Geology* 159: 305–17.

Franck, S., W. von Bloh, C. Bounama, and H.-J. Schellnhuber. 2000. "The Future of the Planet Earth and the Life Span of the Biosphere," in *The Future of the Universe and the Future of Our Civilization*, eds. V. Burdyuzha and G. Khozin. Singapore: World Scientific, pp. 309–15.

Franck, S., A. Block, W. von Bloh, C. Bounama., H.-J. Schellnhuber, and Y. M. Svirezhev. 2000. "Habitable Zone for Earth-like Planets in the Solar System." *Planet. Space Sci.* 48: 1099–105.

Franck, S., A. Block, W. von Bloh, C. Bounama, H.-J. Schellnhuber, and Y. M. Svirezhev. 2000. "Reduction of Life Span as a Consequence of Geodynamics." *Tellus* 52B: 94–107.

Franck, S., W. von Bloh, C. Bounama, M. Steffen, D. Schönberner and H.-J. Schellnhuber. 2001. "Limits of Photosynthesis in Extrasolar Planetary Systems for Earth-like Planets." *Adv. Space Res.* 28/4: 695–700.

Franck, S., A. Block, W. von Bloh, C. Bounama, I. Garrido, and H. J. Schellnhuber. 2001. "Planetary Habitability: Is Earth Commonplace in the Milky Way?" *Naturwiss.* 88: 416–26.

Francois, L. M. and J. C. G. Walker. 1992. "Modelling the Phanerozoic Carbon Cycle and Climate: Constraints from the 87Sr/86Sr Isotopic Ratio of Seawater." *Am. J. Sci.* 292: 81–135.

Freund, R. 1974. "Kinematics of Transform and Transcurrent Faults." *Tectonophysics* 21: 93–134.

Fyfe, W. S. 1978. "The Evolution of the Earth's Crust: Modern Plate Tectonic to Ancient Hotspot Tectonic?" *Chem. Geol.* 23: 89–114.

Gates, D. M. 1993. *Climate Change and Its Biological Consequences.* Sunderland, Mass.: Sinauer Associates.

Gay, K. 1986. *The Greenhouse Effect.* New York: F. Watts.

Glantz, Michael H. 1991. "The Use of Analogies in Forecasting Ecological and Societal Responses to Global Warming" *Environment* 33(5): 10.

Goddéris, Y. and L. M. Francois. 1995. "The Cenozoic Evolution of the Strontium and Carbon Cycle: Relative Importance of Continental Erosion and Mantle Exchanges." *Chem. Geology* 126: 169–90.

Gosse, J. C. et al. 1995. "Precise Cosmogenic 10Be Measurements in Western North America: Support for a Global Younger Dryas Cooling Event." *Geology* 23: 877–80.

Gough, D. O. 1981. "Solar Interior Structure and Luminosity Variations." *Solar Phys.* 74: 21–34.

Gribbin, J. 1990. *Hothouse Earth: The Greenhouse Effect and Gaia.* New York: Grove Weidenfeld.

Gribbin, J. R. 1982. *Future Weather and the Greenhouse Effect.* New York: Delacorte Press/Eleanor Friede.

Grotzinger, J. P., S. A. Bowring, B. Saylor, and A. J. Kauffman. 1995. "New Biostratigraphic and Geochronological Constraints on Early Animal Evolution." *Science* 270: 598–604.

Hare, T. and A. Khan. 1990. *The Greenhouse Effect.* New York: Gloucester Press.

Henderson-Sellers, A. and B. Henderson-Sellers. 1988. "Equable Climate in the Early Archaean." *Nature* 336: 117–18.

Hess, H. H. (1962.) "History of Ocean Basins," in *Petrologic Studies: A Volume to Honor A. F. Buddington,* eds. A. E. J. Engel et al. Boulder, Colo.: Geological Society of America, pp. 599–620.

Hewitt, C. N. and W. T. Sturges. 1993. *Global Atmospheric Chemical Change.* New York: Elsevier Applied Science.

Hoffman, P. F. 1987. "Continental Transform Tectonics: Great Slave Lake Shear Zone (ca. 1.9 Ga), Northwest Canada." *Geology* 15: 785–88.

Hoffman, P. F. 1988. "United Plates of America—the Birth of a Craton." *Ann. Rev. Earth Planet. Sci.* 16: 543–603.

Howell, D. G. and R. W. Murray. 1986. "A Budget for Continental Growth and Denudation." *Science* 233: 446–49.

Hsü, K. J. 1981. "Thin-skinned Plate-tectonic Model for Collision-type Orogenesis." *Sci. Sin.* 24: 100–10.

Jones, P. D. and T. M. L. Wigley. 1990. "Global Warming Trends." *Scientific American* 263(2): 84.

Jorgensen, U. G. 1991. "Advanced Stages in the Evolution of the Sun." *Astron. Astrophysics* 246: 118–36.

Kasting, J. F. 1982. "Stability of Ammonia in the Primitive Terrestrial Atmosphere." *J. Geophys. Res.* 87: 3091–98.

Kasting, J. F. 1984. "Comments on the BLAG Model: The Carbonate-Silicate Geochemical Cycle and Its Effect on Atmospheric Carbon Dioxide over the Past 100 Million Years." *Am. J. Sci.* 284: 1175–82.

Kasting, J. F., J. L. Pollack, and T. P. Ackerman. 1984. "Response of Earth's Atmosphere to Increases in Solar Flux and Implication for Loss of Water from Venus." *Icarus* 57: 335–55.

Kasting, J. F. 1986. "Runaway and Moist Greenhouse Atmospheres and the Evolution of Earth and Venus." *Icarus* 74: 472–94.

Kasting, J. F. 1987. "Theoretical Constraints on Oxygen and Carbon Dioxide Concentrations in the Precambrian Atmosphere." *Precambrian Res.* 34: 205–29.

Kasting, J. F., O. B. Toon, and J. B. Pollack. 1988. "How Climate Evolved on the Terrestrial Planets." *Scientific American* 258,2: 46–53.

Kasting, J. F., D. Whitmire, and R. Reynolds. 1993. "Habitable Zones Around Main Sequence Stars." *Icarus* 101: 108–28.

Kasting, J. F. 1997. "Warming Early Earth and Mars." *Science* 276: 1213–15.

Kellogg, W. W. and R. Schware. 1981. *Climate Change and the Society: Consequences of Increasing Atmospheric Carbon Dioxide.* Boulder, Colo.: Westview Press.

Kirchner, J. W. 1991. "The Gaia Hypotheses: Are They Testable? Are They Useful?" in *Scientists on Gaia,* eds. S. H. Schneider and P. J. Boston. Cambridge, Mass.: M.I.T. Press, pp. 38–46.

Knoll, A. H., A. J. Kaufman, M. A. Semikhatov, J. P. Grotzinger, and

W. Adams. 1995. "Sizing Up the Sub-Tommotian Anconformity in Siberia." *Geology* 23: 1139–43.

Korycansky, D. G., G. Laughlin, and F. C. Adams. 2001. "Astronomical Engineering: A Strategy for Modifying Planetary Orbits." *Astrophysics and Space Science* 275: 349–66.

Krause, F. 1992. *Energy Policy in the Greenhouse*. New York: Wiley.

Kuhn, W. R., J. C. G. Walker, and H. G. Marshall. 1989. "The Affect on Earth's Surface Temperature from Variations in Rotation Rate, Continent Formation, Solar Luminosity, and Carbon Dioxide." *J. Geophys. Res.* 94: 11129–36.

Lasaga, A. C., R. A. Berner, and R. M. Garrels. 1985. "An Improved Geochemical Model of Atmospheric $CO_2$ Fluctuations over Past 100 Million Years," in *The Carbon Cycle and Atmospheric $CO_2$: Natural Variations Archaean to Present*, eds. E. T. Sundquist and W. S. Broecker. Washington, D.C.: Amer. Geophys. Union. pp. 397–411.

Lenton, T. and W. von Bloh. 2001. "Biotic Feedback Extends the Life Span of the Biosphere." *Geophys. Res. Letters* 28: 1715–18.

Lovejoy, T. E. and R. L. Peters. 1992. *Global Warming and Biological Diversity*. New Haven: Yale University Press.

Lovelock, J. E. and L. Margulis. 1974. "Atmospheric Homeostasis by and for the Biophere: The Gaia Hypothesis." *Tellus* 26: 1–10.

Lovelock, J. E. 1979. *Gaia: A New Look at Life on Earth*. Oxford: Oxford University Press.

Lovelock, J. E. and M. Whitfield. 1982. "Life Span of the Biosphere." *Nature* 296: 561–63.

Lovelock, J. E. and A. J. Watson. 1982. "The Regulation of Carbon Dioxide and Climate: Gaia or Geochemistry." *Planet. Space Science* 30: 795–802.

Lovelock, J. E. 1988. *The Ages of Gaia: A Biography of Our Living Earth*. New York: Norton.

Lovelock, J. E. 1992. "A Numerical Model For Biodiversity." *Phil. Trans. R. Soc. London* 338: 383–91.

Marshall, H. G., J. C. G. Walker, and W. R. Kuhn. 1988. "Long-term Climate Change and the Geochemical Cycle of Carbon." *J. Geophys. Res.* 93: 781–801.

Matsui, T. and Y. Abe. 1986. "Evolution of an Impact-induced Atmosphere and Magma Ocean on the Accreting Earth." *Nature* 319: 303–5.

Mayewski, P. A., G. H. Denton, and T. Hughes. 1981. "Late Wisconsin Ice Sheets in North America," in *The Last Great Ice Sheets*, eds. G. H. Denton and T. Hughes. New York: Wiley-Interscience, pp. 67–178.

McCuen, Gary E. 1987. *Our Endangered Atmosphere: Global Warming and the Ozone Layer*. Hudson: Gary E. McCuen Publications.

McElhinny, M. W. 1973, *Paleomagnetism and Plate Tectonics*. Cambridge, England: Cambridge University Press.

McGovern, P. J. and G. Schubert. 1989. "Thermal Evolution of the Earth: Effects of Volatile Exchange between Atmosphere and Interior." *Earth Planet Sci. Lett.* 96: 27–37.

Meissner, R. 1986. *The Continental Crust*. Orlando: Academic Press.

Mesirow, L. E. and S. H. Schneider. 1976. *The Genesis Strategy: Climate and Global Survival*. New York: Plenum Press.

Mitchell, G. J. 1991. *World on Fire: Saving an Endangered Earth*. Toronto: Collier MacMillan.

Muller, R. A. and G. J. MacDonald. 1997. "Simultaneous Presence of Orbital Inclination and Eccentricity in Proxy Climate Records from Ocean Drilling Program Site 806." *Geology* 25: 3–6.

Nance, J. J. 1991. *What Goes Up: The Global Assault on Our Atmosphere*. New York: W. Morrow.

Nesje, A and S. O. Dahl. 2000. *Glaciers and Environmental Change*. London: Arnold.

Newsom, H. E. and J. H. Jones. 1990. *Origin of Earth*. New York: Oxford University Press.

Oppenheimer, M. 1990. *Dead Heat: The Race Against the Greenhouse Effect*. New York: Basic Books.

Owen, T., R. D. Cess, and V. Ramanathan. 1979. "Enhanced $CO_2$ Greenhouse to Compensate for Reduced Solar Luminosity on Early Earth." *Nature* 277: 640–42.

Pachett, J. and C. Chauvel. 1984. "The Mean Life of Continents Is Currently Not Constrained by Nd and Hf Isotopes." *Geophys. Res. Lett.* 11: 151–53.

Raff, R. A. 1996. *The Shape of Life*. Chicago: University of Chicago Press.

Raymo, M. E., W. F. Ruddiman, and P. N. Froelich. 1988. "Influence of Late Cenozoic Mountain Building on Ocean Geochemical Cycles." *Geology* 16: 649–53.

Reckman, A. 1991. *Global Warming*. New York: Gloucester Press.

Reukin, A. 1992. *Global Warming: Understanding the Forecast*. New York: Abbeville Press.

Reymer, A. and G. Schubert. 1984. "Phanerozoic Addition Rates of the Continental Crust and Crustal Growth." *Tectonics* 3: 63–67.

Rybicki, K. R. and C. Denis. 2001. "On the Final Destiny of the Earth and the Solar System." *Icarus* 151: 130–37.

Rye, R., P. H. Kuo, and H. D. Holl. 1997. "Atmospheric Carbon Dioxide Concentrations Before 2.2 Billion Years Ago." *Nature* 378: 603–5.

Sackmann, I. J., A. I. Boothroyd, and K. E. Kraemer. 1993. "Our Sun. III. Present and Future." *The Astrophysical J.* 418: 457–68.

Sagan, C. and G. Mullen. 1972. "Earth and Mars: Evolution of Atmospheres and Surface Temperatures." *Science* 177: 52–56.

Sagan, C. 1977. "Reduced Greenhouse and the Temperature History of the Earth and Mars." *Nature* 269: 224–26.

Sagan, C., A. Druyan, T. Ferris, J. Lomberg, and L. S. Sagan. 1978. *Murmurs of Earth: The Voyager Interstellar Record of Earth*. New York: Random House.

Sagan, C. and C. Chyba. 1997. "The Early Faint Young Sun Paradox: Organic Shielding of Ultraviolet-Labile Greenhouse Gases." *Science* 276: 1217–21.

Schneider, S. 1990. "Prudent Planning for a Warmer Planet." *New Scientist* 128(1743): 49.

Schneider, S. H. 1989. *Global Warming: Are We Entering the Greenhouse Century?* San Francisco: Sierra Club Books.

Schneider, S. H. and P. J. Boston. 1993. *Scientists on Gaia*. Cambridge, Mass.: MIT Press.

Schneider, Stephen H. 1990. "Debating GAIA." *Environment* 32(4): 4.

Schroder, P., R. Smith, and K. Apps. 2001. "Solar Evolution and the Distant Future of Earth." *Astronomy and Geophysics*. 42(6): 26–29.

Schwartzman, D. and S. Shore. 1996. "Biotically Mediated Surface Cooling and Habitability for Complex Life," in *Circumstellar Habitable*

*Zones,* ed. L. Doyle. Menlo Park, Calif.: Travis House Publications, pp. 421–43.

Schwartzman, D. W. and T. Volk. 1989. "Biotic Enhancement of Weathering and the Habitability of Earth." *Nature* 340: 457–60.

Sedjo, R. A. 1989. "Forests: A Tool to Moderate Global Warming." *Environment* 31(1): 14.

Sellers, A. and A. J. Meadows. 1975. "Long-term Variations in the Albedo and Surface Temperature of the Earth." *Nature* 254: 44.

Simons, Paul. 1992. "Why Global Warming Could Take Britain by Storm." *New Scientist* 136(1846): 35.

South, E. L. 1990. *The Changing Atmosphere: A Global Challenge.* New Haven: Yale University Press.

Stevenson, D. J., T. Spohn, and G. Schubert. 1983. "Magnetism and the Thermal Evolution of the Terrestrial Planets." *Icarus* 54: 466–89.

Stommel, H. and E. Stommel. 1979. "The Year Without a Summer." *Scientific American* 240/6: 176–86.

Stumm, W. and J. J. Morgan. 1981. *Aquatic Chemistry.* New York: Wiley-Interscience.

Tajika, E. and T. Matsui. 1992. "Evolution of Terrestrial Proto-$CO_2$ Atmosphere Coupled with Thermal History of the Earth." *Earth Planet. Sci. Lett.* 113: 251–66.

Taylor, S. R. and S. M. McLennan. 1995. "The Geological Evolution of the Continental Crust." *Rev. Geophysics* 33: 241–65.

Tipler, F. J. 2000. "From 2100 to the End of Time." *Wired,* Jan. 2000.

Turcotte, D. L. and G. Schubert. 1982. *Geodynamics.* New York: Wiley.

Uyeda, S. 1987. *The New View of the Earth.* San Francisco: W. H. Freeman and Co.

Valentine, J. W. 1994. "Late Precambrian Bilaterans: Grades and Clades." *Proceedings of the National Academy of Sciences* 91: 6751–57.

Vernadsky, V. 1945. "The Biosphere and the Noosphere." *Amer. Sci.* 33: 1–12.

Vine, F. J. and D. H. Mathews. 1963. "Magnetic Anomalies over Oceanic Ridges." *Nature* 199: 947–49.

Volk, T. 1987. "Feedbacks Between Weathering and Atmospheric $CO_2$ over the Last 100 Million Years." *Am. J. Sci.* 287: 763–79.

Volk, T. 1989. "Rise of Angiosperms as a Factor in Long-term Climatic Cooling." *Geology* 17: 107–10.

Walker, J. C. G., P. B. Hays, and J. F. Kasting. 1981. "A Negative Feedback Mechanism for the Long-term Stabilization of Earth's Surface Temperature." *J. Geophys. Res.* 86: 9776–82.

Walker, J. C. G. 1982. "Climatic Factors on the Archean Earth." *Palaeogeogr. Palaeoclimat. Palaeoecol.* 40: 1–11.

Watson, A. J. and J. E. Lovelock. 1983. "Biological Homeostasis of the Global Environment: the Parable of Daisyworld." *Tellus* 35B: 284–89.

Wegener, A. 1912. "Die Entstehung der Kontinente." *Geol. Rundschau* 3: 276–92.

Wegener, A. 1924. *The Origin of Continents and Oceans.* London: Methuen & Co.

Weidick, A. 1975. "A Review of Quaternary Investigations in Greenland." *Geological Survey of Greenland, Misc. Papers,* pp. 145, 161.

Williams, D. R. and V. Pan. 1992. "Internally Heated Mantle Convection and the Thermal and Degassing History of the Earth." *J. Geophys. Res.* 97 B6: 8937–50.

Wilson, E. 1992. *The Diversity of Life.* Cambridge, Mass.: Harvard University Press.

Wilson, J. T. 1965. "A New Class of Faults and Their Bearing on Continental Drift." *Nature* 207: 343–44.

Wilson, L. and J. W. Head III. 1994. "Mars: Review and Analysis of Volcanic Eruption Theory and Relationships to Observed Landforms." *Rev. Geophys.* 32: 221–63.

Woese, C. R. 1987. "Bacterial Evolution." *Microbiol. Rev.* 51: 221–71.

Woese, C. R., O. Kandler, and M. L. Wheelis. 1990. "Towards a Natural System of Organisms: Proposal for the Domains Archaea, Bacteria, and Eucarya." *Proc. Nat'l. Acad. Sci. USA* 87: 4576–79.

Zahnle, K. J., J. F. Kasting, and J. B. Pollack. 1988. "Evolution of a Steam Atmosphere During Earth's Formation." *Icarus* 74: 62–97.

Zuckerman, B. and M. H. Hart. 1995. *Extraterrestrials: Where Are They?* New York: Cambridge University Press.

## ·ACKNOWLEDGMENTS·

Our exploration of the long-term future of our planet is a logical follow-on from our previous collaboration on the book *Rare Earth*. Together these books examine Earth's ability to support complex life and conclude that our present-day world is rather rare in both space and time. We have tried to the best of our ability to convey the present state of the many scientific endeavors associated with this fascinating but highly interdisciplinary topic. This is a daunting task and we are grateful to the many people that have assisted us along the way. Our main goal was to try to assimilate existing information, the work of many, many people, into a coherent and accurate scientific narrative.

A major inspiration for this book was a symposium that neither of us attended. It took place at the annual meeting of the American Association of Advancement of Science in Washington, D.C., in 2000. The symposium was called "The Far Future Sun and the Ultimate Fate of the Earth" and it was organized by our friend and colleague Lee Anne Willson, an astronomer at Iowa State University. The symposium included talks by Lee Anne on the fates of planets, a talk by Jim Kasting (Pennsylvania State University) on the evolution of Earth's climate, and a talk by Fred Adams

(University of Michigan) entitled "Rest of the Story." This session had only a few talks but they generated broad-ranging interest and served as a catalyst and guide for the present book. We are indebted to Lee Anne for inspiration and insight into several key issues related to the Earth's future.

Many colleagues and friends have helped us with the research and writing of this book. Joseph Kirschvink of the California Institute of Technology read the entire manuscript and aided us with his probing criticism. David Catling and Roger Buick of the University of Washington discussed various scientific matters with us, and Jim Kasting of Pennsylvania State University has helped us with scientific aspects of future modeling. Kevin Zahnle provided comments about a number of intereting topics. The NASA Astrobiology Institute funded much of the research that we discuss in the text. Chris Scotese of the Paleomap Project made the maps of past and future continental positions. The manuscript was materially altered and improved through the help of Bill Dietrich. We want to thank our editor, David Sobel, for his insight and patience, and our agent, Kris Dahl, for shepherding our proposal to its current home.

Numerous faculty members at the University of Washington gave important advice. These included George Wallerstein, Paul Hodge, Bruce Balick, Conway Leovy, Woody Sullivan, Ivan King, Toby Smith, Paula Szkody, Richard Gammon, John Baross, Craig Hogan, and Guillermo Gonzalez, now a faculty member at Iowa State University with Lee Anne Willson.

This book was inspired and abetted by the NASA Astrobiology Institute (NAI), Barry Blumberg, director.

# · I N D E X ·

PETER D. WARD AND DONALD BROWNLEE ARE THE COAUTHORS of the acclaimed and bestselling *Rare Earth*. Don Brownlee is a professor of astronomy at the University of Washington and has been involved in many space experiments and directs *Stardust*, a NASA mission currently flying to a comet. Peter Ward is a professor of geological science and zoology at the University of Washington and the author of nine other books including *Future Evolution*, *In Search of Nautilus*, *The Call of Distant Mammoths*, and *The End of Evolution*, which was a finalist for the *Los Angeles Times* Book Prize.